T0140017

Genetic and Evolutionary Computation

Series Editors

Wolfgang Banzhaf, Department of Computer Science and Engineering, Michigan State University, East Lansing, MI, USA

Kalyanmoy Deb, Department of Electrical and Computer Engineering, Michigan State University, East Lansing, MI, USA

The area of Genetic and Evolutionary Computation has seen an explosion of interest in recent years. Methods based on the variation-selection loop of Darwinian natural evolution have been successfully applied to a whole range of research areas.

The Genetic and Evolutionary Computation Book Series publishes research monographs, edited collections, and graduate-level texts in one of the most exciting areas of Computer Science. As researchers and practitioners alike turn increasingly to search, optimization, and machine-learning methods based on mimicking natural evolution to solve problems across the spectrum of the human endeavor, this growing field will continue to surprise with novel applications and results. Recent award-winning PhD theses, special topics books, workshops and conference proceedings in the areas of EC and Artificial Life Studies are of interest.

Areas of coverage include applications, theoretical foundations, technique extensions and implementation issues of all areas of genetic and evolutionary computation. Topics may include, but are not limited to:

Optimization (multi-objective, multi-level) Design, control, classification, and system identification Data mining and data analytics Pattern recognition and deep learning Evolution in machine learning Evolvable systems of all types Automatic programming and genetic improvement.

Proposals in related fields such as:

Artificial life, artificial chemistries Adaptive behavior and evolutionary robotics Artificial immune systems Agent-based systems Deep neural networks Quantum computing will be considered for publication in this series as long as GEVO techniques are part of or inspiration for the system being described. Manuscripts describing GEVO applications in all areas of engineering, commerce, the sciences, the arts and the humanities are encouraged.

Prospective Authors or Editors:

If you have an idea for a book, we would welcome the opportunity to review your proposal. Should you wish to discuss any potential project further or receive specific information regarding our book proposal requirements, please contact Wolfgang Banzhaf, Kalyan Deb or Mio Sugino:

Areas: Genetic Programming/other Evolutionary Computation Methods, Machine Learning, Artificial Life

Wolfgang Banzhaf Consulting Editor BEACON Center for Evolution in Action Michigan State University, East Lansing, MI 48824 USA banzhafw@msu.edu

Areas: Genetic Algorithms, Optimization, Meta-Heuristics, Engineering

Kalyanmoy Deb Consulting Editor BEACON Center for Evolution in Action Michigan State University, East Lansing, MI 48824 USA kdeb@msu.edu

Mio Sugino mio.sugino@springer.com

The GEVO book series is the result of a merger the two former book series: Genetic Algorithms and Evolutionary Computation https://www.springer.com/series/6008 and Genetic Programming https://www.springer.com/series/6016.

Leonardo Trujillo · Stephan M. Winkler ·
Sara Silva · Wolfgang Banzhaf

Editors

Genetic Programming
Theory and Practice XIX

 Springer

Editors
Leonardo Trujillo 🅾
Engineering Sciences Graduate Program
Tecnológico Nacional de México/Instituto
Tecnológico de Tijuana
Tijuana, Baja California, Mexico

Sara Silva 🅾
Department of Informatics
Faculty of Sciences
University of Lisbon
Lisbon, Portugal

Stephan M. Winkler 🅾
Heuristic and Evolutionary
Algorithms Laboratory
University of Applied Sciences
Upper Austria
Wels, Austria

Wolfgang Banzhaf 🅾
Department of Computer Science
and Engineering
Michigan State University
East Lansing, MI, USA

ISSN 1932-0167   ISSN 1932-0175 (electronic)
Genetic and Evolutionary Computation
ISBN 978-981-19-8462-4   ISBN 978-981-19-8460-0 (eBook)
https://doi.org/10.1007/978-981-19-8460-0

© The Editor(s) (if applicable) and The Author(s), under exclusive license to Springer Nature
Singapore Pte Ltd. 2023
This work is subject to copyright. All rights are solely and exclusively licensed by the Publisher, whether
the whole or part of the material is concerned, specifically the rights of translation, reprinting, reuse
of illustrations, recitation, broadcasting, reproduction on microfilms or in any other physical way, and
transmission or information storage and retrieval, electronic adaptation, computer software, or by similar
or dissimilar methodology now known or hereafter developed.
The use of general descriptive names, registered names, trademarks, service marks, etc. in this publication
does not imply, even in the absence of a specific statement, that such names are exempt from the relevant
protective laws and regulations and therefore free for general use.
The publisher, the authors, and the editors are safe to assume that the advice and information in this book
are believed to be true and accurate at the date of publication. Neither the publisher nor the authors or
the editors give a warranty, expressed or implied, with respect to the material contained herein or for any
errors or omissions that may have been made. The publisher remains neutral with regard to jurisdictional
claims in published maps and institutional affiliations.

This Springer imprint is published by the registered company Springer Nature Singapore Pte Ltd.
The registered company address is: 152 Beach Road, #21-01/04 Gateway East, Singapore 189721,
Singapore

Preface

The COVID-19 pandemic has marked our world indefinitely. Its devastating effects on peoples' lives and livelihoods are, unfortunately for many, incalculable. For the academic and scientific world, it meant a sudden halt to in-person meetings, workshops and conferences, among many other consequences. People had to manage a much greater hardship than having to discuss the latest in science and technology by way of a digital screen, but while most stayed positive and made use of the online opportunities, science, learning and teaching were not the same. After canceling the Genetic Programming Theory and Practice (GPTP) workshop in 2020, we decided to organize the 18th edition of GPTP as an online event in 2021, which allowed us to gather our community and become an active group again. It was a success by any measure, but we still hoped to avoid having to repeat that format again.

For the 19th edition of GPTP in 2022, a mix of hopefulness and nervousness were part of the early organizing meetings, as many institutions around the world were gradually returning, at least partially, to in-person activities. New strains of the SARS-COV-2 virus kept appearing and affected the well-being of many. However, as the first few months of 2022 passed, our hopefulness began to change into joy as we began to realize that holding an in-person event was not impossible, and gradually became a realistic scenario.

Our community was engaged, as people started accepting invitations to the workshop, and booking flights and hotel rooms. It was a strange feeling, since such gatherings are the lifeblood of scientific discourse and engagement. Finally, the time had come to get back to normal, or at least as close to normal as possible: GPTP 2022 was held from June 2 to June 4 at Weiser Hall in the University of Michigan in Ann Arbor, hosted by the Center for the Study of Complex Systems. When Prof. Carl Simon graciously greeted us early on the first day of the event, he reminded us that it was wonderful to host GPTP once again, especially since it was the first in-person event held there in almost two years. While this sounded completely reasonable given what transpired over the past two years, it was still shocking to think about such a statement, and motivated us to make the event as productive and engaging as ever. As Carl mentioned, GPTP is characterized by open discussions, thought-provoking talks and a dynamic format, small in number but big in spirit.

Among the many highlights of GPTP over the years have been the amazing keynote talks, and this year was no exception. The opening keynote was delivered by Dr. Frank Crary from the University of Colorado, Boulder, giving the audience a bird's-eye view, or more appropriately a satellite's-eye view, of how our solar system has been, and is being, explored by robotic spacecraft, particularly by always captivating NASA missions. He also provided a perspective on the role that software systems play in such missions, discussing their unique constraints and stringent requirements, along with the possible opportunities that may lie on the horizon as cost of space travel continues to go down and as more countries reach the frontiers of space, especially for those interested in taking machine learning and genetic programming along for the ride.

On the second day, the keynote was given by Prof. Susan Stepney from the University of York, providing a fresh new look at how life could, or should, be studied, and how to get there by way of an engineering program. At the intersection of biology, computing and physics, cyber-bio-physical systems, or Zoetic systems, were succinctly presented by Susan, capturing the imagination of the audience and discussing how Zoetic science could come to be, by thinking of life as a process, and by taking inspiration of how thermodynamics developed from work in engineering before becoming a full-fledged science. The breadth and scope of the talk was inspiring, and the implications and opportunities for the genetic programming community were discussed in a lively Q&A. These topics are explored in greater detail in the chapter "Life as a Cyber-Bio-physical System" contributed by Susan Stepney to the present collection.

On the final day, the keynote was delivered by Craig Reynolds, who talked about some of his most recent work on the evolution of camouflage using genetic programming and co-evolutionary dynamics. The presentation was extremely engaging, as he showed us how his artificial evolutionary system was able to progressively discover the ability to generate camouflage patterns to trick a learning neural net predator, generating a variety of intriguing and aesthetic visual patterns. From conceptualization to design and implementation, Craig's system presented a perfect example of how researchers in our field continue to discover new and amazing ways to leverage the adaptability of evolution in the systems we develop and study, taking care to provide evolution with the necessary elements to construct the building blocks required to solve a given task.

Besides the keynotes, many invited speakers presented some of their most recent findings and ideas concerning genetic programming, artificial evolution and machine learning, covering topics that included auto-machine learning, interpretable machine learning, adversarial learning, symbolic regression and complexity. The book you hold in your hands contains a collection of 12 chapters derived from all those talks given at the workshop, each chapter having been authored, read, reviewed and discussed at the 19th edition of GPTP in Ann Arbor, Michigan, by participants of the workshop. We also had a fantastic in-person gathering at Bill Worzel's home, hosted by one of the founders of GPTP and his amazing spouse; it was easy to see where the gracious, generous and affable spirit of the event comes from, thanks, Bill!

We are very honored and grateful that we could once again organize another GPTP workshop in person, and the accompanying book, after two years of uncertainty. It is our intention that GPTP continues to be a core event for genetic programming research, bringing together academics, practitioners and theorists from diverse fields of science that intersect in our community, providing for a constructive, thoughtful, inspired and open interchange of ideas, and to do so, whenever possible, in-person, with a coffee during breaks or a beer at dinner.

Tijuana, Baja California, Mexico   Leonardo Trujillo
East Lansing, MI, USA   Wolfgang Banzhaf
Hagenberg, Wels, Austria   Stephan M. Winkler
Lisbon, Portugal   Sara Silva
September 2022

Acknowledgements

We would like to thank all of the participants for making GP Theory and Practice a successful IN-PERSON workshop once again in 2022. It was a breath of fresh air, literally, to talk, interact and discuss with all of the attendees. Our community deserved it after such prolonged isolation. Special thanks to our three wonderful keynote speakers, Frank, Susan and Craig: Your talks were amazing!

We would also like to thank our financial supporters for making the existence of GP Theory and Practice possible for 19 great editions. For 2022, we are grateful to the following sponsors:

- Michael Affenzeller from HEAL and the University of Applied Sciences Upper Austria (FH Oberösterreich)
- Stuart Card
- Michael Korns
- Mark Kotanchek at Evolved Analytics
- John Koza
- Jason H. Moore at the Department of Computational Biomedicine in Cedars-Sinai.

A number of people made key contributions to the organization of the workshop. Foremost among them is Linda Wood, who helped behind the scenes before, during and after the workshop. Special thanks to Carl Simon at the Center for the Study of Complex Systems at the University of Michigan for hosting GPTP once again. We are particularly grateful for contractual assistance by Mio Sugino and Nobuko Kamikawa, Springer-Nature Tokyo, and editorial assistance by Sivananth S. Siva Chandran, Springer-Nature Chennai. We would also like to express our gratitude to

Erik Goodman and Charles Ofria at the BEACON Center for the Study of Evolution in Action at Michigan State University for their continued support.

Tijuana, Baja California, Mexico   Leonardo Trujillo
East Lansing, MI, USA   Wolfgang Banzhaf
Hagenberg, Wels, Austria   Stephan M. Winkler
Lisbon, Portugal   Sara Silva
September 2022

Contents

Symbolic Regression in Materials Science: Discovering Interatomic Potentials from Data 1
Bogdan Burlacu, Michael Kommenda, Gabriel Kronberger,
Stephan M. Winkler, and Michael Affenzeller

Correlation Versus RMSE Loss Functions in Symbolic Regression Tasks 31
Nathan Haut, Wolfgang Banzhaf, and Bill Punch

GUI-Based, Efficient Genetic Programming and AI Planning for Unity3D 57
Robert Gold, Andrew Haydn Grant, Erik Hemberg,
Chathika Gunaratne, and Una-May O'Reilly

Genetic Programming for Interpretable and Explainable Machine Learning 81
Ting Hu

Biological Strategies ParetoGP Enables Analysis of Wide and Ill-Conditioned Data from Nonlinear Systems 91
Mark Kotanchek, Theresa Kotanchek, and Kelvin Kotanchek

GP-Based Generative Adversarial Models 117
Penousal Machado, Francisco Baeta, Tiago Martins, and João Correia

Modeling Hierarchical Architectures with Genetic Programming and Neuroscience Knowledge for Image Classification Through Inferential Knowledge 141
Gustavo Olague, Matthieu Olague, Gerardo Ibarra-Vazquez,
Isnardo Reducindo, Aaron Barrera, Axel Martinez, and Jose Luis Briseño

Life as a Cyber-Bio-Physical System 167
Susan Stepney

STREAMLINE: A Simple, Transparent, End-To-End Automated
Machine Learning Pipeline Facilitating Data Analysis
and Algorithm Comparison 201
Ryan Urbanowicz, Robert Zhang, Yuhan Cui, and Pranshu Suri

Evolving Complexity is Hard 233
Alden H. Wright and Cheyenne L. Laue

ESSAY: Computers Are Useless ... They Only Give Us Answers 255
Bill Worzel

Index 261

Contributors

Michael Affenzeller Heuristic and Evolutionary Algorithms Laboratory, University of Applied Sciences Upper Austria, Hagenberg, Austria

Francisco Baeta CISUC and LASI, Department of Informatics Engineering, University of Coimbra, Coimbra, Portugal

Wolfgang Banzhaf Department of Computer Science and Engineering, Michigan State University, East Lansing, MI, USA

Aaron Barrera EvoVisión Laboratory, CICESE, Ensenada, B.C., Mexico

Jose Luis Briseño EvoVisión Laboratory, CICESE, Ensenada, B.C., Mexico

Bogdan Burlacu Josef Ressel Center for Symbolic Regression and Heuristic and Evolutionary Algorithms Laboratory, University of Applied Sciences Upper Austria, Hagenberg, Austria

João Correia CISUC and LASI, Department of Informatics Engineering, University of Coimbra, Coimbra, Portugal

Yuhan Cui University of Pennsylvania, Philadelphia, PA, USA

Robert Gold ALFA, MIT CSAIL, Cambridge, MA, USA

Andrew Haydn Grant ALFA, MIT CSAIL, Cambridge, MA, USA

Chathika Gunaratne ALFA, MIT CSAIL, Cambridge, MA, USA

Nathan Haut Department of Computational Mathematics, Science and Engineering, Michigan State University, East Lansing, MI, USA

Erik Hemberg ALFA, MIT CSAIL, Cambridge, MA, USA

Ting Hu School of Computing, Queen's University, Kingston, ON, Canada; Department of Computer Science, Memorial University, St. John's, NL, Canada

Gerardo Ibarra-Vazquez ITESM, Institute for Future of Education, Monterrey, N.L., Mexico

Michael Kommenda Josef Ressel Center for Symbolic Regression and Heuristic and Evolutionary Algorithms Laboratory, University of Applied Sciences Upper Austria, Hagenberg, Austria

Kelvin Kotanchek Evolved Analytics LLC, Rancho Santa Fe, CA, USA

Mark Kotanchek Evolved Analytics LLC, Rancho Santa Fe, CA, USA

Theresa Kotanchek Evolved Analytics LLC, Rancho Santa Fe, CA, USA

Gabriel Kronberger Josef Ressel Center for Symbolic Regression and Heuristic and Evolutionary Algorithms Laboratory, University of Applied Sciences Upper Austria, Hagenberg, Austria

Cheyenne L. Laue Computer Science Department, University of Montana, Missoula, USA

Penousal Machado CISUC and LASI, Department of Informatics Engineering, University of Coimbra, Coimbra, Portugal

Axel Martinez EvoVisión Laboratory, CICESE, Ensenada, B.C., Mexico

Tiago Martins CISUC and LASI, Department of Informatics Engineering, University of Coimbra, Coimbra, Portugal

Gustavo Olague EvoVisión Laboratory, CICESE, Ensenada, B.C., Mexico

Matthieu Olague Anahuac University Queretaro, El Márques, Querétaro, Mexico

Una-May O'Reilly ALFA, MIT CSAIL, Cambridge, MA, USA

Bill Punch Department of Computational Mathematics, Science and Engineering, Michigan State University, East Lansing, MI, USA

Isnardo Reducindo Autonomous University of San Luis Potosí, Information Sciences Faculty, Fracc. Talleres, San Luis Potosí, Mexico

Susan Stepney Department of Computer Science, University of York, York, UK

Pranshu Suri University of Pennsylvania, Philadelphia, PA, USA

Ryan Urbanowicz Department of Computational Biomedicine, Cedars Sinai Medical Center, Los Angeles, CA, USA

Stephan M. Winkler Heuristic and Evolutionary Algorithms Laboratory, University of Applied Sciences Upper Austria, Hagenberg, Austria

Bill Worzel Evolution Enterprise, Ann Arbor, MI, USA

Alden H. Wright Computer Science Department, University of Montana, Missoula, USA

Robert Zhang University of Pennsylvania, Philadelphia, PA, USA

Symbolic Regression in Materials Science: Discovering Interatomic Potentials from Data

Bogdan Burlacu, Michael Kommenda, Gabriel Kronberger, Stephan M. Winkler, and Michael Affenzeller

Abstract Particle-based modeling of materials at atomic scale plays an important role in the development of new materials and the understanding of their properties. The accuracy of particle simulations is determined by *interatomic potentials*, which allow calculating the potential energy of an atomic system as a function of atomic coordinates and potentially other properties. First-principles-based *ab initio* potentials can reach arbitrary levels of accuracy, however, their applicability is limited by their high computational cost. Machine learning (ML) has recently emerged as an effective way to offset the high computational costs of *ab initio* atomic potentials by replacing expensive models with highly efficient surrogates trained on electronic structure data. Among a plethora of current methods, symbolic regression (SR) is gaining traction as a powerful "white-box" approach for discovering functional forms of interatomic potentials. This contribution discusses the role of symbolic regression in Materials Science (MS) and offers a comprehensive overview of current methodological challenges and state-of-the-art results. A genetic programming-based approach for modeling atomic potentials from raw data (consisting of snapshots of atomic positions and associated potential energy) is presented and empirically validated on *ab initio* electronic structure data.

B. Burlacu (✉) · M. Kommenda · G. Kronberger
Josef Ressel Center for Symbolic Regression and Heuristic and Evolutionary Algorithms
Laboratory, University of Applied Sciences Upper Austria, Softwarepark 11, 4232 Hagenberg,
Austria
e-mail: bogdan.burlacu@fh-ooe.at

S. M. Winkler · M. Affenzeller
Heuristic and Evolutionary Algorithms Laboratory, University of Applied Sciences Upper
Austria, Softwarepark 11, 4232 Hagenberg, Austria

© The Author(s), under exclusive license to Springer Nature Singapore Pte Ltd. 2023
L. Trujillo et al. (eds.), *Genetic Programming Theory and Practice XIX*,
Genetic and Evolutionary Computation, https://doi.org/10.1007/978-981-19-8460-0_1

1

1 Introduction

Materials Science (MS) is a highly interdisciplinary field incorporating elements of physics, chemistry, engineering and more recently, machine learning , in order to design and discover new materials. The rapid increase in processing power over the last decades has made computational modeling and simulation the main tool for studying new materials and determining their properties and behavior. Computational approaches can deliver accurate quantitative results without the need to set up and execute highly complex and costly physical experiments.

Potential energy surfaces (PES), describing the relationship between an atomic system's potential energy and the geometry of its atoms, are a central concept in computational chemistry and play a pivotal role in particle simulations. An example PES for the water molecule is shown in Fig. 1. The mathematical function used to calculate the potential energy of a system of atoms with given positions in space and generate the PES is called an *interatomic potential* function. The form of this function, it's physical fidelity as well as its complexity and efficiency are critical components in simulations used to predict material properties.

The ability to simulate large particle systems over long time scales depends critically on the accuracy and computational efficiency of the interatomic potential. Broadly speaking, the more accurate the method, the lower its computational efficiency and the more limited its applicability. For example, first-principles modeling methods such as *density functional theory* (DFT) [33] provide highly accurate results by considering quantum-chemical effects but are not efficient enough to simulate large systems containing thousands of atoms over long time scales of nanoseconds [44].

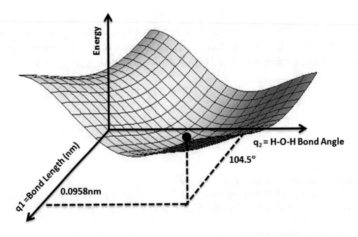

Fig. 1 PES for water molecule: the energy minimum corresponding to optimized molecular structure for water-O-H bond length of 0.0958nm and H-O-H bond angle of 104.5°. Image from Wikipedia ©AimNature

Molecular dynamics (MD) simulations treat materials as systems consisting of many microscopic particles (atoms) which interact with each according to the laws of statistical thermodynamics. These interactions are modeled by interatomic potentials depending mainly on particle positions. Macroscopic properties of materials are obtained as time and/or ensemble averages of processes emerging at the microscopic scale [27].

Empirical and semi-empirical methods treat atomic interactions in a more coarse-grained manner via parameterized analytic functional forms and trade-off accuracy for execution speed in order to enable simulations at a larger scale. Although they are computationally undemanding, they are only able to provide a qualitatively reasonable description of chemical interactions [53].

Machine learning (ML) interatomic potentials aim to bridge the gap between quantum and empirical methods in order to deliver the best of both worlds: functional forms that are as efficient as empirical potentials and as accurate as quantum-chemical approaches.

1.1 Materials Informatics and Data-Driven Potentials

Building upon the three established paradigms of science that have led to many technological advances over time, experimental, theoretical and simulation-based, a fourth "data-driven" paradigm of science is emerging today using machine learning and the large amounts of experimental and simulation data available [1]. "Big-data" science unifies the first three paradigms and opens up new avenues in materials science under the umbrella term of *materials informatics*. The field of material informatics is very new, and many unsolved questions still remain open and wait for proper answers [26].

Machine learning interaction models are generated on the basis of quantum-chemical reference data consisting of a series of snapshots of atomic coordinates, associated potential energy of the system and optionally other properties.

In molecular dynamics simulations, the system's potential energy is typically decomposed into a set of independent m-body interactions that are a function of each particle's position, $\mathbf{r}$. For a two-body or pair potential, it is assumed that the energy contributions from each pair of interacting particles are independent of other pairs and therefore:

$$E = \sum_{\langle i,j \rangle} g(\mathbf{r}_i, \mathbf{r}_j) \tag{1}$$

For a three-body potential, triplets of atoms are also considered:

$$E = \sum_{\langle i,j \rangle} g(\mathbf{r}_i, \mathbf{r}_j) + \sum_{\langle i,j,k \rangle} h(\mathbf{r}_i, \mathbf{r}_j, \mathbf{r}_k) \tag{2}$$

Traditionally, the functions g and h are represented by all kinds of empirical or semi-empirical analytic functions. With the advent of machine learning and data-based modeling, it becomes possible to automatically search for these functional forms with the help of ab initio training data. Substantial effort has already been put into this direction, and many machine learning models have been successful in discovering interatomic potentials for a variety of chemical configurations [42].

1.2 Current Challenges

Despite their success in representing atomic interactions, ML methods are not without their own challenges. Deriving highly accurate and tractable analytic functional forms for high-dimensional PESs is a very active field of research. The most important requirements for ML-based PESs are

- general applicability and absence of ad-hoc approximations (transferability);
- accuracy close to first-principles methods (including high-order many-body effects);
- very high efficiency to enable large simulations;
- the ability to describe chemical reactions and arbitrary atomic configurations;
- the ability to be automatically constructed and systematically improved.

Currently available potentials are far from satisfying all the needs [6], mainly due to the following difficulties and shortcomings.

Physical plausibility
Closed physical systems are governed by various conservation laws that describe invariant properties. These fundamental principles of nature provide strong constraints that can be used to guide the search toward physically plausible ML models [53]. In molecular systems, each conserved quantity is associated with a differentiable symmetry of the action of a physical system. Typical conserved quantities include temporal and roto-translational invariance (i.e. total energy, linear and angular momentum). Forces must be the negative gradient of the potential energy E with respect to atomic positions r_i:

$$F_i = -\nabla r_i E$$

When atoms move, they always acquire the same amount of kinetic energy as they lose in potential energy, and vice versa—the total energy is conserved. The potential energy of a molecule only depends on the relative positions of atoms and does not change with rigid rotations or translations.

Another aspect of invariance is permutational invariance resulting from the fact that from the perspective of the electrons, atoms with the same nuclear charge appear identical to each other and can thus be exchanged without affecting the energy or the forces. To ensure physically meaningful predictions, ML-based models must exhibit the same invariant behavior as the true potential energy surface.

Accuracy

Accuracy is one of the most important requirements of ML potentials. The predicted energies and forces should be as close as possible to the underlying *ab initio* data. Numerical accuracy of the ML models is restricted by the intrinsic limitations of their functional form and descriptors (input variables) used. For example, conceptual problems related to incorporating rotational, translational and permutational invariance into descriptors are of primary relevance [6, 21, 46, 47] as well as their optimal design [20].

Transferability

Ideally, potentials should be generally applicable and should not be restricted to specific types of atomic configurations. Due to their mathematical unbiased form, ML methods are promising candidates to reach this goal. However in practice, developed potentials often perform very well in applications they have been designed for, but are too system-specific and thus cannot be easily transferred from one system to another. The issues of extensibility, generality and transferability of the ML potentials need to be explicitly addressed [6].

Complexity and data requirements

Another issue worth mentioning here is the mathematical complexity of ML potentials. For example, the most popular ML methods used to represent many-body PESs, ANNs, require complex architectures with many adjustable parameters (weights of neural synapses and neuron biases) to yield sufficiently flexible and invariant PES representations . For this, large amounts of training data (often dozens or even hundreds of thousands of points) are needed. On the other hand, the number of training data should be kept as low as possible since they are calculated via demanding quantum-chemical methods. It means that as simple as possible analytic representations of PESs are needed.

Integration of physical knowledge and interpretability

Related to the mathematical complexity issue, it is also important to note that most of the ML methods (e.g. ANN and SVM) are of a "black-box" nature, and may be less amenable to including physical information in the functional forms, relying at least partially on physics-inspired features considered in atomic descriptors. This often leads to the increased mathematical and computational complexity of resulting interaction models. One of the main directions of the current development in ML-based computational MS is the shift from "black-box" methods toward "white-box" methods which often offer better interpretability.

2   State of the Art

A plethora of machine learning approaches have recently emerged as a powerful alternative for finding a functional relation between an atomic configuration and corresponding energy [6, 17, 23]. Several ML techniques such as polynomial fitting [10],

Gaussian processes [5], spectral neighbor analysis [52], modified Shepard interpolation [29], moment tensor potentials [47], interpolating moving least squares [32], support vector machines [4], random forests [31], artificial neural networks (ANNs) [15, 25, 46, 55] or symbolic regression (SR) [40] have been successfully employed for a variety of systems.

More detailed reviews of current ML potentials can be found for example in [23, 37, 42] or [53]. Particularly, ANNs have received considerable attention and are probably the most popular form of ML potentials used in MS [55]. However, methods based on symbolic regression are gaining in popularity due to the advantages they bring in solving aspects of physical knowledge integration, efficiency and interpretability [7, 8, 11, 12, 24, 38, 41, 45, 48]. In the following, we refer to symbolic regression in its canonical incarnation that employs genetic programming to perform a search over the space of mathematical expressions. Symbolic regression approaches have succeeded in rediscovering simple forms of potentials that deliver qualitatively good results in a series of specific applications, some of which are described below.

2.1 Directed Search

The goal of the directed search is to improve search efficiency by limiting the hypothesis space to a functional form known to deliver qualitatively good results, instead of searching for a brand new potential.

Makarov and Metiu [38] use the Morse potential as a functional template for modeling diatomic molecules (see Sect. 5, Eq. 17). They rewrite it in the form:

$$M\big(D(r), R(r)\big) = D(r)\big(1 - \exp\big(R(r)\big)\big)^2 \tag{3}$$

and use genetic programming to find the best $D(r)$ and $R(r)$.

The directed search approach is augmented with an error metric that better reflects the physical characteristics of the problem. A standard error metric such as the MSE has the disadvantage of overemphasizing high-energy points which are rarely used during simulation. For this reason, the authors found it advantageous to introduce a scaling factor:

$$F(a) = \sum_i \frac{\big(E(r_i) - f(r_i; a)\big)^2}{E^2(r_i) + \delta^2} \tag{4}$$

where the constant δ is added to prevent division by zero.

For each function f_α in the population of individuals, the fitness function is then defined as

$$p_\alpha = \exp\big(-\beta F_\alpha\big) \tag{5}$$

where parameter β controls how discriminating the function is and is adaptively updated during the run. The search starts with a small value for β which is gradually increased as the search improves.

The authors note the importance of including the derivative of the energy in the training data:

$$F'(a) = \sum_i \frac{|\nabla E(r_i) - \nabla f(r_i; a)|^2}{|\nabla E(r_i)|^2 + \delta'^2} \tag{6}$$

leading to an expanded fitness function p_α:

$$p_\alpha = \exp\left(-\beta(F + F')\right) \tag{7}$$

The recombination pool is filled using a proportional selection scheme. An additional "natural selection" operator employs a "badness list" $b_\alpha = \exp(\beta F_\alpha)$ whose elements are the inverse of the fitness. Old individuals are replaced with a probability proportional to badness.

Results

The directed search approach is shown to perform better than an undirected search over the search space, on training data generated using the Lippincott potential (Sect. 5, Eq. 19). A population size of 500 individuals is evolved over 150 generations (75,000 evaluations) using the primitive set $\mathscr{P} = \{+, -, \div, \times, \exp\}$. Furthermore, a search directed by a Lennard-Jones potential gives accuracy comparable to that directed by a Morse function, suggesting that restricting the hypothesis space with an appropriate functional template is a powerful and general approach in the search for interatomic potentials. In the case of the Lennard-Jones potential (Sect. 5, Eq. 18), the functional template was defined as

$$f(r) = 4D(r)\left[\frac{1}{4} + \left(\frac{1}{R(r)}\right)^{12} - \left(\frac{1}{R(r)}\right)^6\right] \tag{8}$$

The authors additionally note that some of the returned models, although accurate, exhibited unphysical behavior and did not extrapolate well. For example, one of the returned models based on the Lennard-Jones functional form had very good accuracy but contained a singularity at $r = 12$ Å, a point outside the interpolation range. The authors address overfitting by fitting the parameters of both the energy function and its derivative in the local search phase. This reduces the chance of obtaining pathological curves in the model extrapolation response.

Finally, Makarov and Metiu also model the potential of a triatomic molecule on *ab initio* data consisting of 60 nuclear configurations, showing that directed search maintains high levels of accuracy and scales favorably with dimensionality.

2.2 Directed Search with Parallel Multilevel Genetic Program

Belluci and Coker [7, 8] employ symbolic regression to discover empirical valence bond (EVB) models using directed search augmented with a multilevel genetic programming approach: the lower level (LLGP) optimizes co-evolving populations of models, while the higher level (HLGP) optimizes genetic operator probabilities of the lower level populations. The approach entitled Parallel Multilevel Genetic Program (PMLGP) found accurate EVB models for proton transfer in 3-hydroxy-gramma-pyrone (3-HGP) in the gas phase and protic solvent as well as ultrafast enolketo isomerization in the lowest singlet excited state of 3-hydroxyflavone (3-HF).

At the lower level (LLGP), the authors use the same error metric and fitness as in [38], namely Eqs. 4 and 5. LLGP individuals represent the $R(r)$ functional part of the Morse potential (see Eq. 3). Remarkably, PMLGP does not use crossover but instead uses six different mutation operators:

- *Point mutation* randomly replaces a subtree with a randomly generated one.
- *Branch mutation* replaces a binary operator with one of its arguments at random.
- *Leaf mutation* replaces a leaf node with another randomly selected leaf.
- *New tree mutation* replaces an entire tree with a newly generated tree.
- *Parameter change* replaces each parameter value a_i with $a_i + (R - 0.5)\gamma$, where R is a uniform random number on the unit interval and γ is a scaling constant.
- *Parameter scaling* replaces each parameter value a_i with $a_i R\gamma$, where R is a uniform random number on the unit interval and γ is a scaling constant.

Of the last two types of mutation, parameter change is designed to make small local moves in parameter space, while parameter scaling is designed to make large moves in parameter space to escape the basins of attraction of local optima. Selection is performed using stochastic universal sampling [3].

At the higher level (HLGP), a real vector encoding is used to represent genetic operator probabilities. The population is initialized with k random vectors $P_k = \left(p_1^{(k)}, ..., p_6^{(k)} \right)$, with $\sum_i p_i^{(k)} = 1$, where k ranges from 1 to the total number N_p of processors, such that each vector corresponds to one of the LLGP populations whose operator probabilities it dynamically adapts.

The fitness of each vector P_k is evaluated based on the maximum fitness delta in the corresponding LLGP population over a specified time interval Δt:

$$F_k^{\mathrm{HLGP}} = \frac{\Delta F_{\max}^{\mathrm{LLGP}}}{\Delta t} \qquad (9)$$

This is based on the idea that the larger the magnitude of F_k^{HLGP}, the more successful the set of probabilities P_k at improving the fitness of the population.

Two genetic operators are used to modify the probability vectors P_k:

- *Mutation* changes each component of the vector by a random amount with the constraint that all components sum up to one. This operator kicks in when the fitness of a vector P_k drops below a given threshold.

- *Adaptation* attempts to improve the probability distribution given by P_k by using feedback from the LLGP. Each LLGP builds a histogram of the number of times each mutation produced the most fit member of the population. Then the success frequency of the mutation operator is given by

$$s_i = \frac{w_i m_i}{n}, \quad w_i = \frac{1}{p_i}, \quad n = \sum_i m_i$$

Here, w_i is a weight, m_i is the number of successful mutations for the ith operator (component of P_k) and n is the total number of successful mutations (for all operators). Based on the success frequencies, adaptation shifts a random amount of probability from the least successful operator to the most successful operator.

The number of LLGP populations (and HLGP individuals, respectively) is set to the number of available processors. Initially, all LLGP populations are identical but diverge during evolution as each corresponding fitness function is parameterized with a different value of β evenly sampled over a specified range. In effect, this applies different selection pressures on each LLGP population. Migrations are performed after the last adaptation step in HLGP. At this point, copies of the fittest individual in each LLGP population are sent to all the other populations, where they replace the least fit individual.

Results
Training data for five different diatomic molecules (CO, H_2, HCl, N_2 and O_2) was generated using differently parameterized Morse functions, Gaussian functions and double well functions. The corresponding directed search spaces are given by

$$F_M = D\left(1 - \exp(-R(r; a))\right)^2 + c \qquad \text{Morse} \tag{10}$$

$$F_G = A \exp\left(R(r; a)^2\right) \qquad \text{Gaussian} \tag{11}$$

$$F_D = D_1\left(1 - \exp(-R_1(r; a))\right)^2 + D_2\left(1 - \exp(-R_2(r; a))\right)^2 \quad \text{Double well} \tag{12}$$

Parameters D, c, A, D_1 and D_2 are optimized by including them as leaves in the trees.

The PMLGP approach was compared against a standard parallel genetic programming implementation (SPGP). In both cases, populations of 500 individuals were evolved in parallel on 8 processors for 20,000 generations. The function set $\mathscr{F} = \{+, -, \times, \div, \exp\}$ was used for internal nodes and the terminal set $\mathscr{T} = \{r, a_1, ..., a_{10}\}$ was used for the leaf nodes.

PMLGP was shown to converge faster and achieve higher accuracy than SPGP. The obtained model of the EVB surface accurately reproduced global features of the *ab initio* data. The approach provides a basis for high-quality many-body potentials for studying gas and solution phase photon reactions.

2.3 Parallel Tempering

Slepoy et al. [48] use a hybrid approach consisting of genetic programming, Monte Carlo sampling and parallel tempering to discover the functional form of the Lennard-Jones pair potential.

Parallel tempering is an approach for parallel genetic programming where several islands (or *replicas*) evolve at a different effective temperature. High effective temperatures favor exploration by accepting new trees even if their fitness is poor, and low effective temperatures favor exploitation by being sensitive to small changes in fitness. By using replicas at different temperatures, the approach simultaneously performs both exploitation and exploration.

The remarkable aspect of this approach is that it marks the first large-scale application of genetic programming in materials science with interesting extensions to the canonical Koza-style algorithm and without restrictions of the hypothesis space.

The training data used consists of 10 nuclear configurations of 10 particles placed in 3D space. The Lennard-Jones potential describes the interactions between pairs of particles, therefore a nuclear configuration's energy is given by the sum of pairwise potentials:

$$E_{\text{conf}} = \sum_{<i,j>} V_{\text{LJ}}(r_{ij}) \tag{13}$$

where $r_{ij} = \|\mathbf{r}_i, \mathbf{r}_j\|$ is the distance between particles i and j. Fitness is defined as the negative mean squared error.

The evolutionary search is organized as a three-stage process consisting of generation, mutation and testing. Offspring individuals are tested for acceptance into the new population. A new tree is unconditionally accepted if its fitness exceeds the old one at the same index. Otherwise, it is accepted with the Boltzmann probability:

$$P_{\text{accept}} = \min\left\{1, \exp\left(\frac{F_{\text{new}} - F_{\text{old}}}{T}\right)\right\}$$

where F_{old} and F_{new} are the old and new fitness values, and T is the effective temperature.

After each generation, each sub-population exchanges one tree with its left neighbor in temperature space and one tree with its right neighbor. The trees to be swapped are selected with equal probability from their respective populations. The tree swap is accepted with a probability based on the relative Boltzmann weights of the two trees:

$$P_{\text{acc}} = \min\left\{1, \exp\left[\left(\frac{1}{T_i} - \frac{1}{T_{i+1}}\right)\left(F_{i+1} - F_i\right)\right]\right\}$$

Results
A large-scale experiment was performed on a cluster made of 100 AMD Opteron 2.2 GHz processors. The trees were restricted to minimum depth 3 and maximum depth 4.

200 replicas with temperatures distributed logarithmically from 0.1 to 10 were used. The replica size was chosen to be either $N = 10,000$ or $N = 50,000$ individuals. The primitive set consisted of elementary operations $\mathscr{P} = \{+, -, \times, \div, \exp, | \cdot |\}$.

The proposed approach successfully discovered the Lennard-Jones potential or arithmetic equivalents within 100 generations. Interestingly, the expended effort was estimated to be somewhere in the range of 10^9 evaluated trees, which represents only a small fraction of the possible trees with depth 4 (around 2.9×10^{36}) [48].

A number of ideas for improving the physical fidelity of the developed functional forms and their generality and transferability are suggested:

- Inclusion of additional properties and forces on individual atoms in the training set.
- Primitive set extension to include three-body interactions.
- Integration of physical knowledge (inclusion of symmetries, invariances).

2.4 Symbolically Regressed Table KMC

In order to increase the time scale of simulations, molecular dynamics can be combined with kinetic (dynamic) Monte Carlo (KMC) techniques [9] that coarse-grain the state space, for example via discretization (e.g. assign an atom to a lattice site). The main assumption is that multiscale modeling requires only relevant information at the appropriate length or time scale.

KMC constructs a lookup table consisting of an *a priori* list of events such as atomic jumps or off-lattice jumps. This yields several orders of magnitude increases in simulated time and allows to directly model many processes unapproachable by MD alone. However, identifying barrier energies from a list of events is difficult and restricts the applicability of the method.

Here, symbolic regression is proposed to identify the functional form of the potential energy surface at barrier energy points from a limited set of *ab initio* training data. The method entitled Symbolically Regressed Table KMC (sr-KMC) [45] provides a machine learning replacement for the lookup table in KMC, thus removing the need for explicit calculation of all activation barriers.

Sastry [45] showed that symbolic regression allows atomic-scale information (diffusion barriers on the potential energy surface) to be included in a long-time kinetic simulation without maintaining a detailed description of all atomistic physics, as done within molecular dynamics.

In this approach, fitness is computed as a weighted mean absolute error between the predicted and calculated barriers, for N random configurations:

$$F = \frac{1}{N} \sum_{i=1}^{N} w_i \left| \Delta E_{\text{pred}}(\mathbf{x}_i) - \Delta E_{\text{calc}}(\mathbf{x}_i) \right| \tag{14}$$

Setting $w_i = |\Delta E_{\text{calc}}|^{-1}$ gives preference to predicting accurately lower energy (most significant) events over higher energy events.

The algorithm uses the *ramped-half-and-half* tree creation method, tournament selection and Koza-style subtree crossover, subtree mutation and point mutation [34].

Results

sr-KMC is applied to the problem of vacancy-assisted migration on the surface of phase-separating Cu_xCo_{1-x} at a concentrated alloy composition ($x = 0.5$). Two types of potentials (Morse and TB-SMA) are used to generate the training data via molecular dynamics. The number of active configurations is limited knowing that only atoms in the environment locally around vacancy and migrating atoms significantly influence the barrier energies.

The inline barrier function is represented from the primitive set $\mathscr{P} = \mathscr{F} \cup \mathscr{T}$, with $\mathscr{F} = \{+, -, \times, \div, \text{pow}, \exp, \sin\}$ and $\mathscr{T} = \{\mathbf{x}, \mathscr{R}\}$. Here, $\mathbf{x}$ represents the currently active configuration and $\mathscr{R}$ is an ephemeral random constant.

The results show that GP predicts all barriers within 0.1% error while using less than 3% of the active configurations for training. This leads to a significant scale-up in real simulation time and a significant reduction in the CPU time needed for KMC. sr-KMC is also compared against the basic KMC approach (using a table look up) where it was shown to perform orders of magnitude faster.

The authors note that standard basis-set regression methods are generally not competitive to GP due to the inherent difficulty in choosing appropriate basis functions and show that quadratic and cubic polynomials perform worse in terms of accuracy (within 2.5% error) while requiring energies for $\sim 6\%$ of the active configurations.

They also note that GP is robust to changes in the configuration set, the order in which configurations are used or the labeling scheme used to convert the configuration into a vector of inputs.

2.5 Hierarchical Fair Competition

Brown, Thompson and Schultz [11, 12] are able to rediscover the functional forms of known two- and three-body interatomic potentials using a parallel approach to genetic programming with extensions toward better generalization . Their implementation is based on Hierarchical Fair Competition (HFC) by Jianjun et al. [28].

The HFC framework [28] is designed toward maintaining a continuous supply of fresh genetic diversity in the population and protecting intermediate individuals who have not reached their evolutionary potential from being driven to extinction by unfair competition. It implements these goals with the help of a hierarchical population structure where individuals only compete with other individuals of similar fitness.

Brown et al. note that a correlation -based fitness measure would increase the efficiency of the search and propose the following formula using the Pearson correlation coefficient:

$$F = \cfrac{N}{N + 100 - 100 \left| \displaystyle\sum_{i=1}^{N} \cfrac{(y_i - \bar{y})(\hat{y}_i - \bar{\hat{y}})}{p_i \sigma_y \cdot p_i \sigma_{\hat{y}}} \right|} \tag{15}$$

Here, N is the number of configurations and p_i is the number of terms in the summation over g (see Eq. 1). Ordinary least squares is then used to fit the prediction $\hat{y}$ to the data by introducing scale and intercept terms to the functions g and h:

$$E = \sum_{\langle i,j \rangle} \big(a \cdot g(\mathbf{r}_i, \mathbf{r}_j) + b \big) + \sum_{\langle i,j,k \rangle} \big(c \cdot h(\mathbf{r}_i, \mathbf{r}_j, \mathbf{r}_k) + d \big) \tag{16}$$

The approach is implemented in PM- DREAMER, an open-source software package developed on top of the Open Beagle library for evolutionary computation [19], using its available genetic operators. These include several mutations (standard, shrink, swap, constant), subtree-swapping crossover, tournament selection and elitism:

- *Standard* mutation replaces a node in the tree with a randomly generated subtree.
- *Swap* mutation swaps two nodes in the tree.
- *Shrink* mutation replaces a subtree with one of its arguments.
- *Swap subtree* mutation swaps a subtree's arguments.
- *Ephemeral* mutation changes the value of a constant in the tree.

Additionally, PM- DREAMER implements support for distributed evolution using the MPI standard and introduces migration operators that exchange individuals between sub-populations at fixed intervals.

Bloat reduction strategies are implemented to prevent the expression trees from becoming increasingly large, a tendency observed especially in the case of three-body modeling. Two strategies are tested:

- Using a simplification operator which replaces subtrees that evaluate to a constant value with the constant value: this operator is applied generationally at a fixed interval.
- Using penalty terms to the fitness function: in this case, the fitness is decreased based on a threshold penalty size value s_b and a maximum penalty size s_e, such that trees with length $< s_b$ are not penalized at all, and trees with length $> s_e$ are penalized fully (fitness is set to zero).

Local search. Local search based on the derivative-free Nelder-Mead simplex algorithm is employed with a set probability, optimizing either a single constant or all the constants in the expression.

HFC Extension. Brown et al. implement HFC in a parallel manner by allowing populations with different fitness thresholds to evolve in parallel, with periodic migrations between them. After migrations, populations that grow too large are "decimated" by the removal of the least fit individuals, while populations that grow too small are supplemented with new randomly generated individuals.

The population fitness thresholds are adapted during the search using two strategies: the first strategy uses a percentile parameter p which determines the fitness threshold such that p percent of individuals have equal or lower fitness. The second strategy uses fixed thresholds determined by the first non-zero threshold along with a scaling parameter equal to the ratio between successive thresholds.

Results

Training data for two- and three-body interactions was generated using the Lennard-Jones and Stillinger-Weber potentials (see Sect. 5 Eqs. 18 and 20). In both cases, five configurations were used for training and 50 configurations were used for testing, in order to realistically represent the problem of obtaining models for condensed phases from a small training set. The generated data includes pairwise distances between atoms, the energy and the force on a single atom.

The authors compare a standard parallel island-based evolutionary model against a parallel HFC evolutionary model, using 32 islands with a population of 10 000 individuals each, evolved over a period of 100 generations.

The primitive set used was $\mathscr{P} = \{+, -, \times, \div, \text{pow}, \exp, \log, |\cdot|\}$, tournament selection was used with a tournament size of 6 and, in the case of the standard evolutionary model, 500 individuals were migrated between islands every 5 generations. For the HFC evolutionary model, the migration took place every generation, the first fitness threshold was set to 0.1 and the threshold ratio was set to 1.0. A detailed description of the other algorithm parameters is given in [12].

After an initial tuning phase, the authors note that the number of interactions per energy point greatly increases the runtime requirements for optimization. The C++ implementation of PM- DREAMER is capable of doing a vectorized evaluation of two- and three-body interatomic potential models using SIMD instructions in a manner similar to batched tree interpretation more typically used in GP. With vectorization, the evolutionary algorithm was found to perform roughly four times faster.

Local search was performed with varying probability on all constants in an expression using a maximum of 6 iterations of the Nelder-Mead algorithm. Simplification is performed every 20 generations.

Overall, the authors show that the HFC strategy consistently outperforms the standard generational evolutionary strategy and is able to find very accurate approximations for the targeted empirical potentials (Lennard-Jones and Stillinger-Weber).

2.6 Potential Optimization by Evolutionary Techniques (POET)

POET [24] distinguishes itself from previously described approaches through an extended primitive set which includes summation symbols that aggregate local energy values around each atom, smoothing functions meant to exploit the "short-sightedness" of atomic interactions as well as leaf nodes representing the atomic neighborhood interaction radius.

The primitive set used by the algorithm consists of the function set $\mathscr{F} = \{\sum, f, +, -, \times, \div, \text{pow}\}$ and terminal set $\mathscr{T} = \{\mathscr{R}, r\}$. Here, $\sum$ are summation symbols, f are smoothing symbols, $\mathscr{R}$ represents an ephemeral constant and, like before, r represents the distance between atoms. Distances are considered within the neighborhood of each atom according to inner and outer cutoff radii r_{in} and r_{out}.

An exemplary POET-tree including the special symbols $\sum$, f and r is shown in Fig. 2. This tree corresponds to the following function which returns the predicted value of the local energy E_i around the ith atom considering the distances r_{ij} to its neighbors:

$$E_i = 7.51 \sum_j r_{ij}^{3.98-3.93r_{ij}} f(r_{ij})$$

$$+ \left(28.01 - 0.03 \sum_j r_{ij}^{11.73-2.93r_{ij}} f(r_{ij}) \right) \left(\sum_j f(r_{ij}) \right)^{-1}$$

Hernandez et al. employ a parallel version of genetic programming where twelve populations are evolved simultaneously. The recombination pool in each population is filled from three separate sets of models: a set from the current population, a global set maintained with the overall best (non-dominated) individuals with regard to fitness, complexity and execution speed, and a set of individuals from the other populations. These sets are periodically filled up with individuals at preset intervals.

New individuals are generated by means of crossover and mutation. Crossover replaces a random subtree in the root parent with either another random subtree from another parent or with a linear combination of random subtrees from two different parents. The first method was applied with probability 0.9, while the second method was applied with probability 0.1.

Mutation can replace a subtree with a randomly generated one, swap the arguments of non-commutative symbols or change the symbols of function nodes. Tree initialization is done using Koza's ramped-half-and-half method where the tree depth is sampled from a Gaussian distribution with $\mu = 5$ and $\sigma = 1$.

Local optimization of model coefficients is performed online during the run with the help of a covariance matrix adaptation evolution strategy (CMA-ES) optimizer and a conjugate gradient (CG) optimizer. CMA-ES is used to optimize the coefficients of models in the global set every 10,000 crossover and mutation operations. The CG algorithm is used to perform one optimization step for every newly generated individual.

Results

The proposed approach is validated using training data from DFT molecular dynamics simulations containing snapshots of atomic positions, energies, forces and stresses

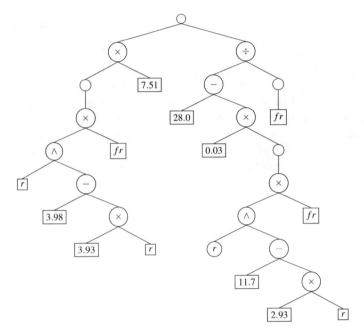

Fig. 2 Interatomic potential representing local energy E_i around the ith atom in electron volts. The expression resembles that of an embedded atom model with a pairwise repulsive term and a many-body attractive term formed by a non-linear transformation of pairwise interactions. Caption and image from Hernandez et al. [24]

for an atomic system of 32 Cu atoms. The fitness measure is an aggregation of the energy, force and stress errors:

$$F = 1000 \cdot \left(0.5\text{MSE}_{\text{energy}} + 0.4\text{MSE}_{\text{force}} + 0.1\text{MSE}_{\text{stress}}\right)$$

The authors demonstrate POET's ability to rediscover Lennard-Jones and Sutton-Chen potentials. The generated models displayed low overfitting and high generalization being able to maintain high predictive accuracy for properties on which they were not trained. The simplicity of the models allows them to predict energies with speeds in the order of microseconds per atom, about 1-4 orders of magnitude faster than other ML potentials. The authors also note that such simple models bring the additional advantage of requiring relatively small amounts of training data.

In terms of runtime performance of POET itself, the authors report 330 CPU-hours spent finding the exact Lennard-Jones potential, 3600 CPU-hours finding the exact Sutton-Chen potential and 360 CPU-hours to find the three best performing models reported in [24]. POET code is open-source and available online.[1]

---

[1] https://gitlab.com/muellergroup/poet.

2.7 Other Applications

Makarov and Metiu [39] use directed genetic programming to find analytic solutions to the time-independent Schrödinger equation. The training data is generated by inverting the Schrödinger equation such that the potential is a functional depending on the wave function and the energy.

Kenoufi and Kholmurodov [30] used symbolic regression to rediscover the Lennard-Jones potential and discovered a new potential for an argon dimer, using *ab initio* data from DFT simulations.

Mueller et al. [41] used symbolic regression for discovering relevant descriptors in hydrogenated nanocrystalline silicon with very low crystalline volume fraction, with applications in improving optical absorption efficiency in thin-film photovoltaics.

Wang et al. [54] used symbolic regression to discover the Johnson-Mehl-Avrami-Kolmogorov transformation kinetics law in the recrystallization process of copper, and the Landau free energy functional form for the displacive tilt transition in perovskite $LaNiO_3$.

Eldridge et al. [18] used the NSGA-III algorithm to learn interatomic potentials for carbon. The approach considered training error, individual age and individual complexity as objectives and was able to find simple and accurate potential functions.

2.8 Summary Discussion

State-of-the-art SR approaches for modeling interatomic potentials recognize the need for domain-specific extensions and hybridizations toward promoting physical plausibility and achieving high accuracy while making the most out of the usually scarce quantities of available *ab initio* training data.

Several extensions and hybridizations are used to augment the classic (Koza-style) genetic programming algorithm and increase its search performance. Parallel, island-based approaches are employed in all of the discussed methods, on the one hand, to more efficiently search the hypothesis space and on the other hand, to achieve higher throughput and alleviate the high computational costs of summations over two- and three-body atomic interactions.

Physical plausibility is promoted by restricting the hypothesis space (directed search), including domain-specific information into the fitness function (e.g. weighing down high-energy points) or including additional targets (forces and stresses).

The results achieved so far have demonstrated the ability of symbolic regression to discover highly accurate and physically plausible functional forms which can increase, due to their simplicity and efficiency, the performance of particle simulations (allowing them to run at larger scales or for longer times). At the same time, since the models are inherently more simple than similar black-box models such as ANNs, they tend to require a lesser amount of training data, which increases their applicability.

Overall, it can be concluded that symbolic regression represents a very promising approach for discovering more accurate and efficient potentials. However, designing evolutionary systems for this application area requires consideration of specific challenges as described in Sect. 1.2. In the following section, we discuss several ideas toward a GP system design which is able to address the domain-specific requirements of interatomic potential.

3 Designing GP for Modeling Interatomic Potentials

The problem of modeling interatomic potentials from data has the main particularity that data comes in the form of atomic configuration snapshots. Each configuration describes the positions of the atoms, the total energy and optionally other properties (forces, stresses). Canonically, these data snapshots are generated by molecular dynamics simulation packages such as LAMMPS [43] or VASP [35] and come in specific formats; see, e.g. POSCAR.[2]

At the minimum, raw data has the following form:

$$\begin{bmatrix} \mathbf{r}_1^{(1)} & \cdots & \mathbf{r}_M^{(1)} & E^{(1)} \\ \vdots & & \vdots & \vdots \\ \mathbf{r}_1^{(N)} & \cdots & \mathbf{r}_M^{(N)} & E^{(N)} \end{bmatrix}$$

where $\mathbf{r}_i^{(k)}$ is the position of the ith atom in the snapshot k and $E^{(k)}$ is the associated potential energy value.

Since atomic interactions are computed based on the distances between atoms, the Cartesian coordinates need to be converted into sets of pairwise distances relative to each atom. It is then the role of the genetic system to evolve an accurate functional relationship between distances r_{ij} and potential energy. For each training sample k, the symbolic regression model needs to process a set of pairwise atomic distances into a prediction for the energy with the help of summation symbol $\sum$.

As it becomes apparent from studying previous approaches described in Sect. 2, modeling interatomic potentials is a non-trivial problem which requires substantial computational resources. Previous implementations employed different strategies for parallelism as well as other optimization techniques such as vectorized model evaluation in order to speed up the search. Additionally, most approaches employed local search in order to improve model coefficients during evolution.

For this reason, we opt to extend the framework Operon [13] with additional functionality for modeling interatomic potentials. Operon already benefits from a fine-grained parallelism model designed for scalability and was shown to perform well on a variety of symbolic regression problems [14]. Additionally, it features sup-

---

[2] https://www.vasp.at/wiki/index.php/POSCAR.

Taskflow: NSGA2

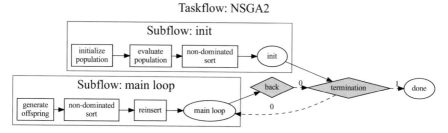

Fig. 3 Taskflow describing the NSGA2 algorithm in Operon. Each individual task within the subflows (initialization, evaluation, offspring generation) executes in parallel, using a number of logical threads equal to the population size

port for local optimization using the Levenberg-Marquardt optimization algorithm, where the gradient is obtained via automatic differentiation.

We adopt a multi-objective approach based on the NSGA2 algorithm [16] where model length is used alongside prediction accuracy in order to promote parsimony, interpretability and generalization.

3.1 Symbolic Regression in Operon

Operon is a C++ framework for symbolic regression that employs logical parallelism during evolution, such that every new offspring individual is generated in its own logical thread. An example evolutionary algorithm implemented in Operon as an operator graph is shown in Fig. 3.

Operon uses a variable-length linear encoding where each tree is represented as a postfix sequence of nodes. Each node has typical attributes such as length, depth, arity and opcode. Evaluation efficiency is achieved by employing a batched tree interpreter, which iterates over the tree nodes and executes the corresponding functions on fixed-size batches of data. As the batch size is known at compile-time, these operations are vectorized. The entire tree evaluation infrastructure relies on the *Eigen* C++ library [22] for efficient, vectorized execution.

3.1.1 Implementing the $\sum$ Symbol

The tree interpreter represents a generic approach to tree evaluation and is agnostic of the actual primitive set used by the algorithm. Each node is mapped to a callable[3] (stateful function object) that defines the functional transformation. The callables themselves are required to satisfy a certain function signature and to operate in both scalar and dual number domains.

---

[3] https://en.wikipedia.org/wiki/Callable_object.

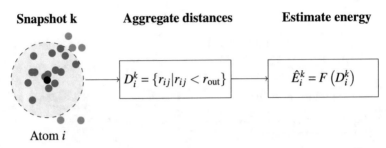

Fig. 4 Prediction of atom energies using SR. The total energy is then $\hat{E}^{(k)} = \sum_i \hat{E}_i^{(k)}$

This mechanism facilitates the extension of the default primitive set with any kind of ad hoc functionality—the $\sum$ (summation) symbol in this particular application. Figure 4 shows the general workflow for processing a set of atomic positions into pairwise distances and using them to estimate the potential energy. The function F represents a symbolic functional form which includes $\sum$ symbols over pairwise atomic distances. Since the $\sum$ symbol is essentially a reduction operator,[4] the actual number of dimensions of the input data is three: N snapshots, M atoms and L pairwise distances (where L dynamically depends on the cutoff radius r_{out}).

Like many other evolutionary frameworks, Operon relies on a dataset object which holds tabular data in two dimensions: X observations $\times$ Y features. Therefore, it is not straightforward to accommodate an additional data dimension without significant redesign work. However, it is relatively easy to incorporate an extra, inner dataset into the function object associated with the $\sum$ symbol, which will contain the values in the third dimension (interatomic distances).

For this mechanism to work, a convention is necessary: the *outer* dataset will contain the target energy values as well as an input variable r whose value is always 1 (this value was chosen arbitrarily as a non-problematic constant which does not cause discontinuities). The variable r simply acts as a placeholder for the pairwise atomic distance values. The *inner* dataset will contain the actual pairwise distances under the same input name r. The distances are computed when the atomic coordinate values are loaded into the callable. A nested tree interpreter is then used to evaluate the current $\sum$-subtree using the inner dataset as input. Similar to Hernandez et al. [24], the $\sum$ symbol also applies a smoothing function on its output (see Eq. (7) in [24]) with the inner and outer cutoff radii equal to 3 Å and 5 Å, respectively.

Under this set of rules, a leaf node corresponding to the input variable r will evaluate to 1 when not under a $\sum$ symbol, and to the set of pairwise distance values corresponding to the current atomic configuration otherwise. Additionally, to disallow nesting of $\sum$ symbols, the behavior is dynamically switched depending on the surrounding tree context: if a $\sum$ symbol finds itself under another $\sum$ symbol, then it simply acts as the identity function $f(x) = x$. This convention does not impact the evolutionary system's ability to discover interatomic potential functional forms.

---

[4] https://en.wikipedia.org/wiki/Reduction_operator.

3.2 Empirical Validation

We demonstrate the capabilities of the proposed NSGA2-based multi-objective approach using the *ab initio* data used by Hernandez et al. [24]. This data consists of 150 snapshots of 32-atom DFT molecular dynamics simulations of copper (Cu): 50 snapshots at 300 K (NVT), 50 snapshots at 1400 K (NVT) and 50 snapshots at 1400 K (NPT at 100 kPa). Although the data also contains components of forces and virial stress tensors, only the energy was used as a modeling target in this experiment. The data consisting of 150 configurations is shuffled and split equally into training and test partitions.

3.2.1 Experimental Setup

The experiment used a fixed set of parameters shown in Table 1. The primitive set was varied and consisted of different symbol combinations, as shown in Table 2: with and without the power function, and alternating between the plain division symbol $\div$ and the analytic quotient: $aq(x) = \sqrt{x^2 + 1}$. The population size was chosen to be 10,000 based on the assumption that a larger population size can increase the coverage of the search space and thus produce better results. Following the first batch of results, a subsequent sensitivity test with smaller population sizes ($N = \{1000, 5000\}$) revealed that our initial choice of population size caused the algorithm to produce statistically significant better results on the training data, but not on the test data, where all population sizes performed equally well.

Two input variables are used: r as a placeholder for atomic distances and $q = \frac{1}{r}$ as a placeholder for the inverse of r, given that some empirical potentials like Lennard-Jones explicitly use the inverse in their formula. Each experimental configuration was repeated 50 times and the median values were reported (with the exception of runtime, which was averaged). Errors were reported as median $\pm$ standard deviation. Model length was computed as the length of the simplified representation returned by *Sympy*, using the infix textual representation of the best individual as input.

3.2.2 Results

Results aggregated over 50 runs for each configuration are shown in Table 2, alongside p-value matrices computed using the Kruskal-Wallis H-test [36]. Significance is encoded in Tables 3 and 4 using font weight and color: values lower than $\alpha = 0.01$ are shown in bold black font, values lower than $\alpha = 0.05$ are shown in black font, while all the other values are shown in gray. The direction of the relationship is determined using a comparison of median values and shown as $\uparrow$ (worse/higher error) or $\downarrow$ (better/lower error) symbols prefixed to the values.

The overall best models from all runs and all configurations are shown in Table 5. These models have been selected based on both test accuracy and simplicity of their

Table 1 NSGA2 parameters

Population size	10,000 individuals
Tree initialization	Balanced tree creator (BTC) [13]
Max tree length	20
Max tree depth	10
Crossover probability	100%
Crossover operator	Subtree crossover
Mutation probability	25%
Mutation operator	Uniformly chosen from
	• Subtree removal/insertion/replacement
	• Change function symbol
	• Change variable name
	• Additive one point leaf mutation $(v = v + \mathcal{N}(0, 1))$
	• Discrete point leaf mutation ($v \leftarrow$ math constant: π, e, ...)
Selection operator	Crowded tournament selection [16], group size = 17
Objectives	Pearson R^2 and model length
Evaluation budget	10^8 Fitness evaluations

functional form. Two other models with better test scores have been discarded due to very complex structures or very large coefficient values. Table 5 illustrates this fact by displaying the absolute rank of each model (based purely on test accuracy and disregarding other criteria).

Interestingly, the arithmetic-only configurations A, B and C generated 4 out of 5 of the selected best models. Although configuration A produced significantly worse ($p < 0.01$) training accuracies than all other configurations, it did not produce worse models in terms of generalization, where it is only worse than E. Nevertheless, the explicit inclusion of $1/r$ as an input seems to help the search.

It is also worth noting that configurations using the analytic quotient instead of (unprotected) division generally perform better on the training data ($p < 0.05$), but do not perform better on the test data. For example, configuration K is better than A, B, C, H and I in terms of training accuracy, but is not better than any of them in terms of the test (on the contrary, it is worse than E at $p < 0.05$). From this, we can surmise that in this particular test setting and for this particular data, AQ does not offer an advantage compared to normal division.

Table 2 Operon NSGA2 Results

ID	Primitive set	Inputs	MAE$_{train}$	MAE$_{test}$	Length	Runtime (s)
A	$\sum, +, -, \times, \div$	r	0.568 ± 0.045	0.602 ± 0.059	32.0	118.52
B	$\sum, +, -, \times, \div$	q	0.518 ± 0.036	0.599 ± 0.069	44.0	142.01
C	$\sum, +, -, \times, \div$	r, q	0.512 ± 0.043	0.595 ± 0.091	42.0	143.69
D	$\sum, +, -, \times, aq$	r	0.498 ± 0.047	0.583 ± 0.060	56.0	165.49
E	$\sum, +, -, \times, aq$	q	0.500 ± 0.066	0.574 ± 0.068	56.5	162.51
F	$\sum, +, -, \times, aq$	r, q	0.493 ± 0.046	0.593 ± 0.060	60.0	169.64
G	$\sum, +, -, \times, \div, pow$	r	0.501 ± 0.042	0.620 ± 0.039	39.0	286.95
H	$\sum, +, -, \times, \div, pow$	q	0.516 ± 0.048	0.604 ± 0.065	46.5	241.25
I	$\sum, +, -, \times, \div, pow$	r, q	0.514 ± 0.051	0.596 ± 0.057	47.0	290.53
J	$\sum, +, -, \times, aq, pow$	r	0.507 ± 0.052	0.608 ± 0.059	47.0	269.26
K	$\sum, +, -, \times, aq, pow$	q	0.489 ± 0.053	0.623 ± 0.085	57.0	244.44
L	$\sum, +, -, \times, aq, pow$	r, q	0.497 ± 0.053	0.594 ± 0.068	57.0	281.86

Overall, judging from median error values and statistical significance p-values, there is no clear winner among the tested configurations. However, a pattern emerges when observing the functional forms of the best models, mostly originating from configurations B and C. After simplification using *Sympy*, the models become highly similar with the same mathematical structure consisting of a sum of three factors in the numerator (each including the inverse of r) and another sum in the denominator (also including the inverse of r). Although these models are remarkably simple, further testing is required to validate their properties and behavior.

In terms of runtime, the proposed approach is efficient, with the longest run taking on average 290 seconds to evolve a population of 10,000 individuals for 1000 generations on a single multicore computer. In comparison, Hernandez et al. [24] report 360 CPU-hours expended on finding accurate GP models.

Code availability. The C++ implementation, data and results are freely available online at https://github.com/foolnotion/atomic-potentials.

Table 3 Training error p-value matrix using the Kruskal statistical test. Significance shown by bold black font ($p < 0.01$), black font ($p < 0.05$) and gray (no significance). Relationship direction given by comparison of medians: ↑ (worse/higher error), ↓ (better/lower error)

	A	B	C	D	E	F	G	H	I	J	K	L
A		↑**4e-07**	↑**8e-07**	↑**1e-09**	↑**2e-07**	↑**1e-09**	↑**1e-08**	↑**3e-06**	↑**8e-06**	↑**3e-07**	↑**9e-10**	↑**9e-09**
B	↓**4e-07**		↑6e-01	↑2e-02	↑1e-01	↑2e-02	↑1e-01	↑1e+00	↑8e-01	↑3e-01	↑**6e-03**	↑5e-02
C	↓**8e-07**	↓6e-01		↑7e-02	↑2e-01	↑5e-02	↑3e-01	↓7e-01	↓9e-01	↑5e-01	↑2e-02	↑1e-01
D	↓**1e-09**	↓2e-02	↓7e-02		↓4e-01	↑9e-01	↓3e-01	↓4e-02	↓6e-02	↓3e-01	↑7e-01	↑8e-01
E	↓**2e-07**	↓1e-01	↓2e-01	↑4e-01		↑5e-01	↓8e-01	↓1e-01	↓3e-01	↓9e-01	↑2e-01	↑7e-01
F	↓**1e-09**	↓2e-02	↓5e-02	↓9e-01	↓5e-01		↓4e-01	↓2e-02	↓6e-02	↓3e-01	↑6e-01	↓8e-01
G	↓**1e-08**	↓1e-01	↓3e-01	↑3e-01	↑8e-01	↑4e-01		↓1e-01	↓3e-01	↓8e-01	↑2e-01	↑6e-01
H	↓**3e-06**	↓1e+00	↑7e-01	↑4e-02	↑1e-01	↑2e-02	↑1e-01		↑8e-01	↑3e-01	↑**7e-03**	↑5e-02
I	↓**8e-06**	↓8e-01	↑9e-01	↑6e-02	↑3e-01	↑6e-02	↓3e-01	↓8e-01		↑4e-01	↑3e-02	↑1e-01
J	↓**3e-07**	↓3e-01	↓5e-01	↑3e-01	↓9e-01	↑3e-01	↑8e-01	↓3e-01	↓4e-01		↑1e-01	↑5e-01
K	↓**9e-10**	↓**6e-03**	↓**2e-02**	↓7e-01	↓2e-01	↓6e-01	↓2e-01	↓**7e-03**	↓3e-02	↓1e-01		↓4e-01
L	↓**9e-09**	↓5e-02	↓1e-01	↓8e-01	↓7e-01	↑8e-01	↓6e-01	↓5e-02	↓1e-01	↓5e-01	↑4e-01	

4 Conclusion

This work surveyed the main applications of SR in Materials Science, namely for the discovery of simple and efficient models of interatomic potentials. Both previous results, as well as results obtained by our own proposed approach and described in this paper, suggest that SR is capable of finding accurate models that can further the capabilities of particle simulations.

Similar to POET [24], our approach does not restrict the search space in any way (with the exception of tree length and depth limits) and is therefore capable of finding models that do not resemble previously known, empirical potential functions. At the same time, should a directed search be required, the framework is trivial to extend with this feature.

Empirical testing shows that relatively simple primitive sets are powerful enough to discover accurate potential functions with good extrapolation behavior. On this data, no advantage was found in using the analytic quotient over standard division. More experiments will be required to establish the benefits of larger primitive sets, for example ones that include logarithmic, exponential or trigonometric functions.

Table 4 Test error p-value matrix using the Kruskal statistical test. Significance shown by bold black font ($p < 0.01$), black font ($p < 0.05$) and gray (no significance). Relationship direction given by comparison of medians: ↑ (worse/higher error), ↓ (better/lower error)

	A	B	C	D	E	F	G	H	I	J	K	L
A		↑3e-01	↑4e-01	↑2e-01	↑**7e-03**	↑8e-02	↓9e-01	↓4e-01	↑1e-01	↓5e-01	↓9e-01	↑7e-02
B	↓3e-01		↑9e-01	↑1e+00	↑2e-01	↑6e-01	↓2e-01	↓7e-01	↑6e-01	↓8e-02	↓2e-01	↑4e-01
C	↓4e-01	↓9e-01		↑1e+00	↑2e-01	↑5e-01	↓4e-01	↓8e-01	↓8e-01	↓1e-01	↓3e-01	↑6e-01
D	↓2e-01	↓1e+00	↓1e+00		↑1e-01	↓6e-01	↓3e-01	↓8e-01	↓9e-01	↓4e-02	↓3e-01	↓6e-01
E	↓**7e-03**	↓2e-01	↓2e-01	↓1e-01		↓4e-01	↑**3e-03**	↓8e-02	↓3e-01	↓**1e-03**	↓1e-02	↓4e-01
F	↓8e-02	↓6e-01	↓5e-01	↓6e-01	↑4e-01		↓5e-02	↓4e-01	↓7e-01	↓1e-02	↓8e-02	↓9e-01
G	↑9e-01	↑2e-01	↑4e-01	↑3e-01	↑**3e-03**	↑5e-02		↑3e-01	↑3e-02	↑5e-01	↓7e-01	↑2e-02
H	↑4e-01	↑7e-01	↑8e-01	↑8e-01	↑8e-02	↑4e-01	↓3e-01		↑4e-01	↓1e-01	↓3e-01	↑3e-01
I	↓1e-01	↓6e-01	↓8e-01	↓9e-01	↑3e-01	↑7e-01	↓3e-02	↓4e-01		↓2e-02	↓1e-01	↑7e-01
J	↑5e-01	↑8e-02	↑1e-01	↑4e-02	↑**1e-03**	↑1e-02	↓5e-01	↑1e-01	↑2e-02		↓7e-01	↑**7e-03**
K	↑9e-01	↑2e-01	↑3e-01	↑3e-01	↑1e-02	↑8e-02	↑7e-01	↑3e-01	↑1e-01	↑7e-01		↑7e-02
L	↓7e-02	↓4e-01	↓6e-01	↓6e-01	↑4e-01	↑9e-01	↓2e-02	↓3e-01	↓7e-01	↓**7e-03**	↓7e-02	

Several other aspects like a more comprehensive search in the space of hyper-parameters or an exploration of the effects of local search also need to be fully investigated in the future. Compared to other works described in our survey, our approach did not diverge from the "vanilla" version of GP, using a classical multi-objective approach (NSGA2) together with a domain-specific primitive set. It will be also worthwhile to explore various ways to scale up the search using multiple populations and more sophisticated evolutionary models.

Future development directions include expanding the capabilities of the framework to include three- or many-body interactions, considering model derivatives in order to model atomic forces as well, and overall improving its ability to incorporate and respect the fundamental laws of this kind of physical system.

Table 5 Overall best models, where ID identifies the configuration in Table 2

ID	Model
C	$MAE_{train} = 0.579$, $MAE_{test} = 0.448$, Absolute rank: 1
	$$-110.531 - \frac{2929.411 \sum \left(\left(-0.974 + \frac{2.68}{r}\right)\left(0.727 - \frac{2.888}{r}\right)\left(0.727 - \frac{1.747}{r}\right)\right)}{\sum\left(-\frac{0.972}{r(0.899r - 1.815)}\right)}$$
E	$MAE_{train} = 0.612$, $MAE_{test} = 0.454$, Absolute rank: 2
	$$\frac{3.037\left(-\sum\left(\frac{0.211}{r^2}\right) - 2.396\right)}{\sqrt{\sum^2\left(\left(-2.409 + \frac{6.254}{r}\right)\left(1.209 - \frac{4.99}{r}\right)\left(1.209 - \frac{2.956}{r}\right)\right) + 1}} - 101.086$$
C	$MAE_{train} = 0.585$, $MAE_{test} = 0.458$, Absolute rank: 3
	$$-111.611 + \frac{12327.356\sum\left(\left(-0.817 + \frac{2.014}{r}\right)\left(0.318 - \frac{1.255}{r}\right)\left(0.706 - \frac{1.913}{r}\right)\right)}{\sum\left(\frac{0.806}{r\left(0.307r - \frac{1.292}{r}\right)}\right)}$$
C	$MAE_{train} = 0.550$, $MAE_{test} = 0.473$, Absolute rank: 6
	$$-108.409 + \frac{14618.749\sum\left(\left(0.555 - \frac{1.538}{r}\right)\left(0.707 - \frac{1.667}{r}\right)\left(0.787 - \frac{3.116}{r}\right)\right)}{\sum\left(\frac{3.142}{r(1.922 - 0.953r)}\right)}$$
B	$MAE_{train} = 0.549$, $MAE_{test} = 0.475$, Absolute rank: 7
	$$-109.903 - \frac{82734.094\sum\left(\left(-0.361 + \frac{1.414}{r}\right)\left(0.527 - \frac{1.433}{r}\right)\left(0.622 - \frac{1.512}{r}\right)\right)}{\sum\left(\frac{0.873}{r\left(-0.339 + \frac{0.686}{r}\right)}\right)}$$

5 Appendix

Empirical potentials

For a comprehensive overview of empirical potentials, we recommend the work of Araújo and Ballester [2]. Below, we give a casual overview of the most important empirical potentials mentioned in this contribution.

Morse potential

This is an empirical potential used to model diatomic molecules:

$$V_M(r) = D\left(1 - \exp\left(-a(r - r_0)\right)^2\right) \tag{17}$$

where D is the dissociation energy, r is the distance between atoms, a is a set of parameters and r_0 is the equilibrium bond distance.

Lennard-Jones potential

The Lennard-Jones potential models soft repulsive and attractive interactions and can describe electronically neutral atoms or molecules. Interacting particles repel each other at very close distances, attract each other at moderate distances, and do not interact at infinite distances:

$$V_{\text{LJ}}(r) = 4\varepsilon \left[\left(\frac{\sigma}{r} \right)^{12} - \left(\frac{\sigma}{r} \right)^{6} \right] \tag{18}$$

where r is the distance between atoms, ε is the dispersion energy and σ is the distance at which the particle-particle potential energy V is zero.

Lippincott potential

Lippincott [49] potential involves an exponential of interatomic distances

$$V_{\text{LIP}}(r) = D \left(1 - \exp \left(\frac{-n(r - r_0)^2}{2r} \right) \right) \left(1 + aF(r) \right) \tag{19}$$

where D is the dissociation energy, r is the distance between atoms, r_0 is the equilibrium bond distance and a and n are parameters. $F(r)$ is a function of internuclear distance such that $F(r) = 0$ when $r = \infty$ and $F(r) = \infty$ when $r = 0$.

Stillinger-Weber potential

The Stillinger-Weber potential [50] models two- and three-body interactions by taking into account not only the distances between atoms but also the bond angles:

$$V_{\text{SW}}(r) = \sum_{\langle i,j \rangle} \phi_2(r_{ij}) + \sum_{\langle i,j,k \rangle} \phi_3(r_{ij}, r_{ik}, \theta_{ijk}) \tag{20}$$

where

$$\phi_2(r_{ij}) = A\varepsilon \left[B \left(\frac{\sigma}{r_{ij}} \right)^p - \left(\frac{\sigma}{r_{ij}} \right)^q \right] \exp \left(\frac{\sigma}{r_{ij} - a\sigma} \right) \text{ and} \tag{21}$$

$$\phi_3(r_{ij}, r_{ik}, \theta_{ijk}) = \lambda\varepsilon \left[\cos\theta_{ijk} - \cos\theta_0 \right]^2 \times \exp \left(\frac{\gamma\sigma}{r_{ij} - a\sigma} \right) \exp \left(\frac{\gamma\sigma}{r_{ik} - a\sigma} \right) \tag{22}$$

Sutton-Chen potential

The Sutton-Chen potential [51] has been used in molecular dynamics and Monte Carlo simulations of metallic systems. It offers a reasonable description of various bulk properties, with an approximate many-body representation of the delocalized metallic bonding:

$$V_{\text{SC}} = \sum_{\langle i,j \rangle} U(r_{ij}) - \sum_i u\sqrt{\rho_i} \tag{23}$$

Here, the first term represents the repulsion between atomic cores, and the second term models the bonding energy due to the electrons. Both terms are further defined in terms of reciprocal power so that the complete expression is

$$V_{\text{SC}} = \epsilon \left[\sum_{\langle i,j \rangle} \left(\frac{a}{r_{ij}} \right)^n - C \sum_i \sqrt{\sum_j \left(\frac{a}{r_{ij}} \right)^m} \right] \tag{24}$$

where C is a dimensionless parameter, ϵ is a parameter with dimensions of energy, a is the lattice constant, m, n are positive integers with $n > m$ and r_{ij} is the distance between the ith and jth atoms.

References

1. Agrawal, A., Choudhary, A.: Perspective: Materials informatics and big data: realization of the "fourth paradigm" of science in materials science. APL Mater. **4**(5), 053208 (2016)
2. Araújo, J.P., Ballester, M.Y.: A comparative review of 50 analytical representation of potential energy interaction for diatomic systems: 100 years of history. Int. J. Quantum Chem. **121**(24), e26808 (2021)
3. Baker, J.E.: Reducing bias and inefficiency in the selection algorithm. In: Proceedings of the Second International Conference on Genetic Algorithms on Genetic Algorithms and Their Application, pp. 14–21, L. Erlbaum Associates Inc., USA (1987)
4. Balabin, R.M., Lomakina, E.I.: Support vector machine regression (ls-svm)–an alternative to artificial neural networks (anns) for the analysis of quantum chemistry data? Phys. Chem. Chem. Phys. **13**, 11710–11718 (2011)
5. Bartók, A.P., Kondor, R., Csányi, G.: On representing chemical environments. Phys. Rev. B **87**, 184115 (2013)
6. Behler, J.: Perspective: Machine learning potentials for atomistic simulations. J. Chem. Phys. **145**(17), 170901 (2016)
7. Bellucci, M.A., Coker, D.F.: Empirical valence bond models for reactive potential energy surfaces: A parallel multilevel genetic program approach. J. Chem. Phys. **135**(4), 044115 (2011)
8. Bellucci, M.A., Coker, D.F.: Molecular dynamics of excited state intramolecular proton transfer: 3-hydroxyflavone in solution. J. Chem. Phys. **136**(19), 194505 (2012)
9. Binder, K., Heermann, D., Roelofs, L., John Mallinckrodt, A., McKay, S.: Monte carlo simulation in statistical physics. Comput. Phys. **7**(2), 156–157 (1993)
10. Brown, A., McCoy, A.B., Braams, B.J., Jin, Z., Bowman, J.M.: Quantum and classical studies of vibrational motion of ch5+ on a global potential energy surface obtained from a novel ab initio direct dynamics approach. J. Chem. Phys. **121**(9), 4105–4116 (2004)
11. Brown, M.W., Thompson, A.P., Watson, J.-P., Schultz, P.A.: Bridging scales from ab initio models to predictive empirical models for complex materials. Technical report, Laboratories, Sandia National (2008)
12. Brown, W.M., Thompson, A.P., Schultz, P.A.: Efficient hybrid evolutionary optimization of interatomic potential models. J. Chem. Phys. **132**(2), 024108 (2010)
13. Burlacu, B., Kronberger, G., Kommenda, M.: Operon C++: an efficient genetic programming framework for symbolic regression. In: Proceedings of the 2020 Genetic and Evolutionary Computation Conference Companion, GECCO'20, pp. 1562–1570. Association for Computing Machinery (2020). (internet, 8–12 July 2020)
14. La Cava, W.G., Orzechowski, P., Burlacu, B., de França, F.O., Virgolin, M., Jin, Y., Kommenda, M., Moore, J.H.: Contemporary symbolic regression methods and their relative performance (2021). CoRR, arXiv:2107.14351
15. Chen, R., Shao, K., Fu, B., Zhang, D.H.: Fitting potential energy surfaces with fundamental invariant neural network. ii. generating fundamental invariants for molecular systems with up to ten atoms. J. Chem. Phys. **152**(20), 204307 (2020)
16. Deb, K., Agrawal, S., Pratap, A., Meyarivan, T.: A fast and elitist multiobjective genetic algorithm: Nsga-ii. IEEE Trans. Evol. Comput. **6**(2), 182–197 (2002)
17. Dral, P.O.: Quantum chemistry in the age of machine learning. J. Phys. Chem. Lett. **11**(6), 2336–2347 (2020). PMID: 32125858

18. Eldridge, A., Rodriguez, A., Hu, M., Hu, J.: Genetic programming-based learning of carbon interatomic potential for materials discovery (2022)
19. Gagné, C., Parizeau, M.: Genericity in evolutionary computation software tools: Principles and case study. Int. J. Artif. Intell. Tools **15**(2), 173–194 (2006)
20. Gao, H., Wang, J., Sun, J.: Improve the performance of machine-learning potentials by optimizing descriptors. J. Chem. Phys. **150**(24), 244110 (2019)
21. Ghiringhelli, L.M., Vybiral, J., Levchenko, S.V., Draxl, C., Scheffler, M.: Big data of materials science: Critical role of the descriptor. Phys. Rev. Lett. **114**, 105503 (2015)
22. Guennebaud, G., Jacob, B., et al.: Eigen v3 (2010). http://eigen.tuxfamily.org
23. Handley, C.M., Behler, J.: Next generation interatomic potentials for condensed systems. Eur. Phys. J. B **87**(7), 152 (2014)
24. Hernandez, A., Balasubramanian, A., Yuan, F., Mason, S.A.M., Mueller, T.: Fast, accurate, and transferable many-body interatomic potentials by symbolic regression. NPJ Comput. Mater. **5**(1), 112 (2019)
25. Hey, T., Butler, K., Jackson, S., Thiyagalingam, J.: Machine learning and big scientific data. Philos. Trans. R. Soc. A Math. Phys. Eng. Sci. **378**(2166), 20190054 (2020)
26. Himanen, L., Geurts, A., Foster, A.S., Rinke, P.: Data-driven materials science: status, challenges, and perspectives. Adv. Sci. **6**(21), 1900808 (2019)
27. Hospital, A., Goñi, J.R., Orozco, M., Gelpí, J.L.: Molecular dynamics simulations: advances and applications. Adv. Appl. Bioinform. Chem. AABC **8**, 37 (2015)
28. Hu, J., Goodman, E., Seo, K., Fan, Z., Rosenberg, R.: The hierarchical fair competition (hfc) framework for sustainable evolutionary algorithms. Evol. Comput. **13**(2), 241–277 (06 2005)
29. Ischtwan, J., Collins, M.A.: Molecular potential energy surfaces by interpolation. J. Chem. Phys. **100**(11), 8080–8088 (1994)
30. Kenoufi, A., Kholmurodov, K.: Symbolic regression of interatomic potentials via genetic programming. Biol. Chem. Res **2**, 1–10 (2015)
31. Kim, C., Pilania, G., Ramprasad, R.: From organized high-throughput data to phenomenological theory using machine learning: the example of dielectric breakdown. Chem. Mater. **28**(5), 1304–1311 (2016)
32. Kim, K.H., Lee, Y.S., Ishida, T., Jeung, G.-H.: Dynamics calculations for the lih+h li+h2 reactions using interpolations of accurate ab initio potential energy surfaces. J. Chem. Phys. **119**(9), 4689–4693 (2003)
33. Kohn, W., Sham, L.J.: Self-consistent equations including exchange and correlation effects. Phys. Rev. **140**, A1133–A1138 (1965)
34. Koza, J.R.: Genetic Programming: On the Programming of Computers by Means of Natural Selection. MIT Press, Cambridge, MA, USA (1992)
35. Kresse, G., Furthmüller, J.: Efficient iterative schemes for ab initio total-energy calculations using a plane-wave basis set. Phys. Rev. B **54**, 11169–11186 (1996)
36. Kruskal, W.H., Allen Wallis, W.: Use of ranks in one-criterion variance analysis. J. Am. Stat. Assoc. **47**(260), 583–621 (1952)
37. Kusne, A., Mueller, T., Ramprasad, R.: Machine learning in materials science: recent progress and emerging applications. Rev. Comput. Chem. (2016). (2016-05-06)
38. Makarov, D.E., Metiu, H.: Fitting potential-energy surfaces: a search in the function space by directed genetic programming. J. Chem. Phys. **108**(2), 590–598 (1998)
39. Makarov, D.E., Metiu, H.: Using genetic programming to solve the schrödinger equation. J. Phys. Chem. A **104**(37), 8540–8545 (2000)
40. Mueller, T., Hernandez, A., Wang, C.: Machine learning for interatomic potential models. J. Chem. Phys. **152**(5), 050902 (2020)
41. Mueller, T., Johlin, E., Grossman, J.C.: Origins of hole traps in hydrogenated nanocrystalline and amorphous silicon revealed through machine learning. Phys. Rev. B **89**, 115202 (2014)
42. Pilania, G.: Machine learning in materials science: From explainable predictions to autonomous design. Comput. Mater. Sci. **193**, 110360 (2021)
43. Plimpton, S.: Fast parallel algorithms for short-range molecular dynamics. J. Comput. Phys. **117**(1), 1–19 (1995)

44. Rothe, T., Schuster, J., Teichert, F., Lorenz, E.E.: Machine Learning Potentials-State of the Research and Potential Applications for Carbon Nanostructures. Technische Universität, Faculty of Natural Sciences, Institute of Physics (2019)
45. Sastry, K.N.: Genetic algorithms and genetic programming for multiscale modeling: Applications in materials science and chemistry and advances in scalability. PhD thesis, University of Illinois, Urbana-Champaign (March 2007)
46. Shao, K., Chen, J., Zhao, Z., Zhang, D.H.: Communication: fitting potential energy surfaces with fundamental invariant neural network. J. Chem. Phys. **145**(7), 071101 (2016)
47. Shapeev, A.V.: Moment tensor potentials: a class of systematically improvable interatomic potentials. Multiscale Model. Simul. **14**(3), 1153–1173 (2016)
48. Slepoy, A., Peters, M.D., Thompson, A.P.: Searching for globally optimal functional forms for interatomic potentials using genetic programming with parallel tempering. J. Comput. Chem. **28**(15), 2465–2471 (2007)
49. Steele, D., Lippincott, E.R., Vanderslice, J.T.: Comparative study of empirical internuclear potential functions. Rev. Mod. Phys. **34**, 239–251 (1962)
50. Stillinger, F.H., Weber, T.A.: Computer simulation of local order in condensed phases of silicon. Phys. Rev. B **31**, 5262–5271 (1985)
51. Sutton, A.P., Chen, J.: Long-range finnis-sinclair potentials. Philos. Mag. Lett. **61**(3), 139–146 (1990)
52. Thompson, A.P., Swiler, L.P., Trott, C.R., Foiles, S.M., Tucker, G.J.: Spectral neighbor analysis method for automated generation of quantum-accurate interatomic potentials. J. Comput. Phys. **285**, 316–330 (2015)
53. Unke, O.T., Chmiela, S., Sauceda, H.E., Gastegger, M., Poltavsky, I., Schütt, K.T., Tkatchenko, A., Müller, K.-R.: Machine learning force fields. Chem. Rev. **0**(0):null. PMID: 33705118 (2021)
54. Wang, Y., Wagner, N., Rondinelli, J.M.: Symbolic regression in materials science. MRS Commun. **9**(3), 793–805 (2019)
55. Zhang, L., Han, J., Wang, H., Car, R., Weinan, E.: Deep potential molecular dynamics: a scalable model with the accuracy of quantum mechanics. Phys. Rev. Lett. 143001 (2018)

Correlation Versus RMSE Loss Functions in Symbolic Regression Tasks

Nathan Haut, Wolfgang Banzhaf, and Bill Punch

Abstract The use of correlation as a fitness function is explored in symbolic regression tasks and its performance is compared against a more typical RMSE fitness function. Using correlation with an alignment step to conclude the evolution led to significant performance gains over RMSE as a fitness function. Employing correlation as a fitness function led to solutions being found in fewer generations compared to RMSE. We also found that fewer data points were needed in a training set to discover correct equations. The Feynman Symbolic Regression Benchmark as well as several other old and recent GP benchmark problems were used to evaluate performance.

1   Introduction

Symbolic Regression (SR) has long been a hallmark of Genetic Programming applications. Already in John Koza's first book on GP it features prominently among the problems of program induction that he mentions as problems Genetic Programming can attempt to solve, on in the first table on page 15 [1]. Among the many applications listed in that table, optimal control, empirical discovery and forecasting, symbolic

N. Haut (✉) · B. Punch
Department of Computational Mathematics, Science and Engineering, Michigan State University, East Lansing, MI, USA
e-mail: hautnath@msu.edu

B. Punch
e-mail: punch@msu.edu

W. Banzhaf
Department of Computer Science and Engineering, Michigan State University, East Lansing, MI, USA
e-mail: banzhafw@msu.edu

© The Author(s), under exclusive license to Springer Nature Singapore Pte Ltd. 2023
L. Trujillo et al. (eds.), *Genetic Programming Theory and Practice XIX*,
Genetic and Evolutionary Computation, https://doi.org/10.1007/978-981-19-8460-0_2

integration or differentiation, inverse problems, and discovering mathematical identities could be easily coached in terms of symbolic regression. To cite the author: "*[...], symbolic regression involves finding a model that fits a given sample of data*". The model is constructed from mathematical functions and their numeric coefficients "*that provides a good, best or perfect fit*".

In the absence of substantial progress on the automatic programming front in the early years of GP, symbolic regression and its Machine Learning (ML) cousin of pattern classification have taken on extremely important roles in Genetic Programming. In fact, regression and classification are now often used as *the* classical application examples of Genetic Programming, recognizing the contributions GP has made to those fields in the preceding decades.

The textbook on Genetic Programming by one of the authors of this chapter [2] argued strongly in favor of the view that GP can be considered part of the Machine Learning field. To apply GP to those tasks seemed to us at the time a low-hanging fruit, and the success and widespread use of GP in these applications today proves that point. However, the second and probably more ambitious goal of GP is *automatic programming* in the general sense, which comprises program synthesis, program repair, probably program analysis, etc. We pointed out this ambition at the time, but did not see how to proceed until evolutionary and genetic improvement [3, 4] came around. These two techniques have now thoroughly paved the way for the more general goal of GP, but it remains to be seen what results can be harvested from these techniques and re-invigorated approaches of GP to automatic programming.

Turning back to symbolic regression, Affenzeller et al. [5] explain the key task of SR as finding a [symbolic, mathematical] relationship between a dependent variable y (output) and a set of specified independent (input) variables $\mathbf{x}$

$$y = f(\mathbf{x}, \mathbf{a}) + \varepsilon \tag{1}$$

with f the functional relationship, $\mathbf{a}$ some coefficients to modify the functional structure, and ε a noise term. Here we have taken the liberty to modify the authors equation by introducing a vector notation for inputs and coefficients (note that these vectors have different dimensionality). We would also like to emphasize that the additive noise is but one type of possible noise in the system, as well as that the task could comprise the finding of relationships for more than one output variable y, thus rendering both y and f as vectors. Both input and target values are normally given as data points, perhaps produced by measurements of a system, with the expectation that the SR algorithm produces a mathematical model that is able to reproduce those data points and both interpolate among them and extrapolate beyond them.

In recent years, a number of different non-GP symbolic regression methods have been proposed. One such method is the physics-inspired method by Udrescu and Tegmark [6] which made use of physical knowledge to restrict the search space of model creation in order to arrive at solutions in reasonable time. These authors also proposed collections of 100 new benchmark datasets (including a set of further *bonus* datasets) based on the known relationship of physical quantities and correspondingly derived data. Among the physical knowledge injected into the search process were considerations of unit dimensionality, translational symmetry, and multiplicative separability of the resulting models.

What struck us was the efficiency with which certain equations could be found, sometimes even with 10 data points. At the same time, we realized that the physical inspiration, while a strength in terms of knowledge injection into the process, was an ad hoc solution that could not be transferred easily to systems of an unknown type. It seemed to us that the emphasis on global features of a model (rather than a point-by-point comparison of data points) was hinted in these results. We also knew that researchers had used other fitness functions with good outcomes, notably the Pearson correlation coefficient.

In this chapter, we report on experiments with GP symbolic regression where the fitness function, the traditional loss function called root mean square error (RMSE) is given by

$$L = \sqrt{\frac{1}{N} \sum_{i=1}^{N} (y_i - \hat{y}_i)^2}, \tag{2}$$

where N is the number of data points i, y_i is the target output, and $\hat{y}_i$, the output calculated by the program under consideration, is replaced by the correlation function

$$R = \frac{\sum_{i=1}^{N} (y_i - \bar{y})(\hat{y}_i - \bar{\hat{y}})}{\sqrt{\sum_{i=1}^{N} (y_i - \bar{y})^2 \times \sum_{i=1}^{N} (\hat{y}_i - \bar{\hat{y}})^2}} \tag{3}$$

of target versus program output.

This replacement is, however, not direct. First off, we try to maximize R^2 (or minimize $1 - R^2$), but then, in a post-processing step, we align the resulting relationship via a simple linear regression step, minimizing

$$\underset{a_0, a_1}{\mathrm{argmin}} \sum_{i=1}^{N} (|y_i - (a_1 \hat{y}_i + a_0)|). \tag{4}$$

The essential difference between these two fitness functions is the *global* consideration the subtracted averages of Eq. (3) bring in. You can note that they are in relation to their respective output data series (target or program). They are thus entering information about the *shape* of the entire curve into the fitness function while the absolute scaling and translation is left to the linear regression post-processing step. We could say that the correlation function looks at the relative position of data points in the target dataset and compares that to the relative position in the program/model produced dataset.

Correlation functions have been used in the past in genetic programming and in other data analysis applications, but often for sorting out the independence and therefore relevance of input features. Thus, it was used to identify dependent features from the input and therefore fight the curse of dimensionality, certainly a legitimate application. But applying correlation as a fitness function to compare target and program output yielded results on the AI Feynman benchmark datasets that were surprising to us. In fact, when we submitted some results to GECCO 2022, reviewers were incredulous and rejected the manuscript as a full paper. Frequently, the results showed that three data points are sufficient to come up with the correct equation or relationship between data points.

The book chapter presented here will examine the performance of the correlation fitness function in more detail and compare it to the RMSE loss function along a number of axes—number of data points required, noise level, and dimensionality of the input.

2 Related Work

In [7], the authors develop a new fitting method relying on the maximization of the correlation coefficient between two sets of data that could well be random or systematically related. The correlation coefficient between two random variables X and Y

$$R = \frac{cov(X, Y)}{\sigma_X \sigma_Y}, \tag{5}$$

where X, Y take on values $X_1, ..., X_N$ and $Y_1, ..., Y_N$ is a well-defined quantity confined to the range $[-1, +1]$. If R is close to either 1 or -1, the data are very strongly correlated to each other, if R is close to 0, there is virtually no correlation between the two series. By minimizing

$$1 - R^2 \tag{6}$$

the goal of this optimization is the same as the more familiar mean square error minimization. Using a continuous example (functions instead of data points and integrals instead of sums) the authors can derive that the correlation-based fitting method is orders of magnitude less sensitive than the MSE method. In other words, finding a fit is much easier using the correlation-based method.

Expressed differently, by searching for the maximum correlation between two curves/sets of points, we allow for an infinite set of possible solutions (irrespective of translation and scaling), while by searching for the minimum of MSE, we allow only one solution to be possible. More recent work by one of the same authors [8] generalizes their sensitivity analysis to an entire family of fitting methods that are based on different L_q norms.

Keijzer [9], revisited by [10], lays out an example of a symbolic regression problem where the issue of (R)MSE becomes clearly visible. A simple target function

$$y = x^2 \tag{7}$$

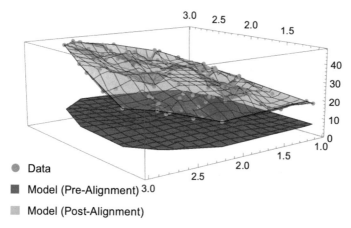

Fig. 1 Correlation as a fitness function can identify successful models that would not be identified by RMSE. The red surface represents a model that was identified as good by the correlation fitness function and the green surface is the same model after has been aligned. Orange points represent the raw data, which consists of 100 points. Both the model and the aligned model have an R^2 value of 0.9999. The pre-aligned model has an RMSE value of 22.24 and the aligned model has an RMSE value of $2.076 * 10^{-14}$

and a modified target function

$$y = x^2 + 100 \tag{8}$$

are compared in terms of their performance under symbolic regression. The authors comment that the simple addition of a constant was able to mislead their GP systems in the search space, resulting in only 16% of their runs (with given representation and parameters) being successful, compared to the unmodified regression problem, which was solved by 98% of their runs. Keijzer suggests a linear scaling method, among other measures like interval arithmetic, to address such problems. While useful under special circumstances, the simple replacement by the MSE fitness function with a correlation fitness function (plus its post-processing step) discussed here perfectly solves this problem.

Figure 1 demonstrates how correlation as a fitness function is capable of identifying potentially good models that RMSE would not identify as a good model. This is a result of correlation assigning fitness independent of a linear scaling and shift, so it can identify a model that has the correct shape but may not yet be in the correct location or at the correct scale.

3 Benchmarks

A number of different benchmarks have been proposed for symbolic regression tasks. GP started out with the symbolic regression problems used in Koza's first book [1]. Over time some more complex problems were added culminating in the suggestion by White et al. [11] to consider these as new standard benchmarks. Korns [12–14] has given a series of presentations at GPTP on systematically more difficult symbolic regression problems and their solution, increasingly relying on hybrid algorithms to solve them. Finally, Udrescu and Tegmark have proposed a collection of physical laws extracted from Feynman lectures [6] as a good benchmark suite for symbolic regression algorithms.

In the following, we shall briefly discuss the former, giving examples of each, before focusing on the latter, the so-called AI Feynman benchmark set.

3.1 Koza's Benchmarks

Koza was the first to highlight the joys and sorrows of symbolic regression with Genetic Programming. The quartic polynomial

$$x^4 + x^3 + x^2 + x \tag{9}$$

was the first discussed in [1], though one could argue that the bang-bang control problem discussed in an earlier section was close. This problem later proliferated to the following problems:

$$x^4 - x^3 + x^2 - x \tag{10}$$

$$x^4 + 2x^3 + 3x^2 + 4x \tag{11}$$

$$x^6 - 2x^4 + x^2. \tag{12}$$

Many more problems of different types were proposed in the book (e.g., symbolic derivatives and symbolic integration, equivalence relationships, roots of equations, etc.).

3.2 New Benchmark Standards

Keijzer extended the benchmark set studied in [9] to the Keijzer instances which were further expanded by Vladislavleva et al. [15] and Nguyen et al. [16]. Typical examples are Keijzer-5:

$$\frac{30xz}{(x - 10)y^2} \tag{13}$$

Vladislavleva-1:

$$\frac{e^{-(x_1 - 1)^2}}{1.2 + (x_2 - 2.5)^2} \tag{14}$$

or Nguyen-5:

$$sin(x^2)cos(x) - 1 \tag{15}$$

just to name a few.

A 2012 community survey [17] revealed the mainly used benchmarks and was summarized and standardized in [11].

3.3 The GPTP Benchmarks

While the term GPTP benchmarks is actually broader, we would like to focus here on the series of contributions and problem suggestions by Korns and his co-authors [12–14, 18–22].

Since this is a large set of problems, we are going to select only one here, Korns-8:

$$6.87 + 11\sqrt{7.23x_0x_3x_4} = 6.87 + 29.58\sqrt{x_0x_3x_4} \tag{16}$$

out of five dimensions $x_0, ..., x_4$, where some variables (x_1, x_2) do not carry information but only noise.

3.4 Feynman Symbolic Regression Benchmark

Here we shall mainly focus on the AI Feynman set of equations/data, lifted out of the lectures of Richard Feynman [23–25].

4 Methods

Symbolic regression was performed using StackGP, a stack-based genetic programming system. The parameters chosen for the system are shown in Table 1. It is important to note that the two sub-populations are evolved in parallel yet do not interact until completion. Upon completion, the two populations are merged and the most fit individuals in the combined population are then selected and returned as the final population of a run.

To compare the performance of using RMSE against correlation as a fitness function, we explored how noise, number of points, and dimensionality affect the resulting fitnesses of the best individuals found during evolution.

For each problem and set of conditions (noise and number of points), a total of 100 repeated independent trials were conducted and the median fitness of the best models from each trial was computed using the test data for the associated problem.

Table 1 Evolution parameter settings

Parameter	Setting
Mutation rate	79
Crossover rate	11
Spawn rate	10
Elitism rate	10
Crossover method	2-point
Tournament size	5
Population size	300
Independent runs	100
Sub-populations	2
Termination criteria	2 Min. (wall time)

To make for a simple comparison, both models trained using RMSE and correlation as their fitness functions were evaluated using RMSE on the test data. The test data consisted of 200 points generated without noise added to determine how close the evolved models are to the true generating equation.

4.1 Noise Introduction

Uniformly distributed multiplicative random noise was introduced to the response data $y = f(x)$ by supplying a percentage to a noise generating function:

$$y = f(x)(1 + \varepsilon), \tag{17}$$

where

$$\varepsilon \in \left[-\frac{R}{2}, \frac{R}{2} \right] \tag{18}$$

is a uniformly distributed random variable from the interval $[-\frac{R}{2}, \frac{R}{2}]$.

4.2 Varying Number of Points

For most problems, between 3 and 193 points were used to determine how changing the number of data points affects the success of the search. This was performed by initially testing with 3 points, then adding 20 points until a total of 193 points were tested. For some of the Feynman problems, the number of points varied from 3 to 19 points incrementing by 2 to observe how small changes in the number of points impacts method performance. In each independent repeated trial, the points used were generated randomly anew.

4.3 Dimensional Sensitivity

Sensitivity to dimensionality was explored by observing the variation in success between the different problems which vary from 1 dimension up to 9 dimensions.

4.4 Sensitivity to Constants

The sensitivity to identifying equations correctly when constants are introduced was explored by introducing constants of varying magnitude and determining how the error is affected by the magnitude of constants.

5 Results

5.1 The Keijzer-5 Benchmark

The results of comparing the correlation-based fitness function to the usual RMSE on the new benchmark standards with 20–200 points and 10% noise are shown in Fig. 2. The results show that correlation finds more accurate models than RMSE with 10% noise for the Keijzer-5 benchmark problem.

In Fig. 3, noise sensitivity was explored with 2,000 points with noise between 0 and 20% on Keijzer-5. While the correlation approach is sensitive to multiplicative noise and gradually deteriorates as the amplitude of noise is increased, the correlation approach generally finds more accurate models even with noise as high as 20%.

5.2 The Korns-8 Benchmark

Figure 4 now compares the results of runs using correlation and RMSE as fitness functions to try to solve the Korns-8 benchmark problem, ranging from 3 to 193 training data points with no (0%) noise. The correlation-based fitness function consistently finds essentially perfect solutions with more than 3 points, while the RMSE-based fitness function performed relatively poorly for all numbers of data points, with only a slight improvement as the number of points increases.

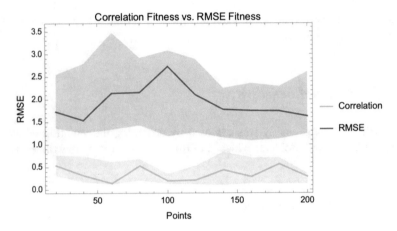

Fig. 2 Comparing using RMSE and correlation as the fitness function on Keijzer-5 with 10% noise

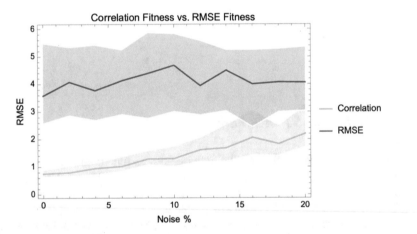

Fig. 3 Comparing using RMSE and correlation as the fitness function on Keijzer-5 with varying noise to determine noise tolerance when using 2,000 training points

The noise tolerance of correlation and RMSE was also explored on Korns-8 by varying noise from 0 to 20% using 200 and 2,000 training points. The results are shown in Figs. 5 and 6, respectively. With 200 data points (Fig. 5) the correlation approach stops outperforming RMSE when 12% or more noise is included in the data. When the data has 2,000 training points, however, as shown in Fig. 6, we see that while the correlation approach shows sensitivity to the amount of noise present, it still finds better models with data that has 20% noise, demonstrating that more data can effectively counter noise with a correlation fitness function.

What is interesting to note is that when comparing the R^2 values between the two approaches with using 2,000 points, it can be seen in Fig. 7 that R^2 values for all models found using correlation are 1, up to 20% noise. This indicates that the correct

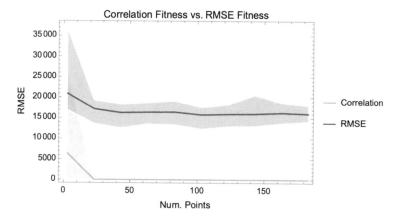

Fig. 4 Comparing using RMSE and correlation as the fitness function on Korns-8 with 0% noise

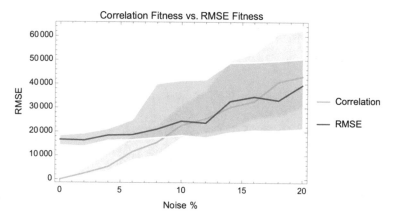

Fig. 5 Comparing using RMSE and correlation as the fitness function on Korns-8 as noise increases from 0 to 20% with 200 training points. With 12% noise or more the correlation approach becomes comparable or worse than RMSE

model form was still found in all cases and points to the alignment step as the part that is sensitive to noise. An example solution is shown in Eq. (19), on comparing to original function (Eq. 16):

$$949.216 + 21.536\sqrt{x_0 x_3 x_4}. \qquad (19)$$

When looking at the R^2 values from the 200 point cases, we can see there is a very clear threshold where the signal-to-noise ratio becomes too small and the quality of models drops off significantly by using the correlation approach. The behavior of the correlation fitness (green dots) in Fig. 8 looks like a phase transition at around 10% noise. The functional relationship evolved before the transition (e.g., at 2% noise) is still correct:

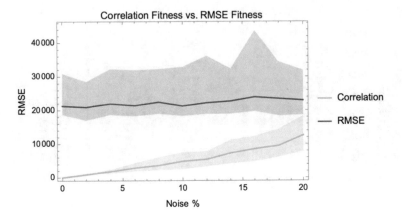

Fig. 6 Comparing using RMSE and correlation as the fitness function on Korns-8 as noise increases from 0 to 20% with 2000 training points

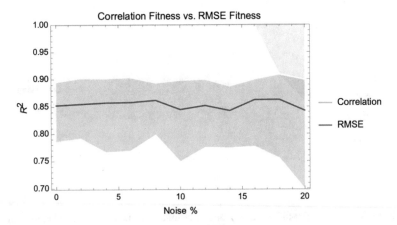

Fig. 7 Comparing using RMSE and correlation as the fitness function on Korns-8 as noise increases from 0 to 20% with 2,000 training points using R^2 as the metric for comparison

$$564.468 + 23.315\sqrt{x_0 x_3 x_4}. \tag{20}$$

However, with 14% noise (after the transition) the functional relationship is not any more correct with:

$$1406.43 + 0.133 x_0 x_3 x_4. \tag{21}$$

The correct variables are still found, but the square root function is not any more present.

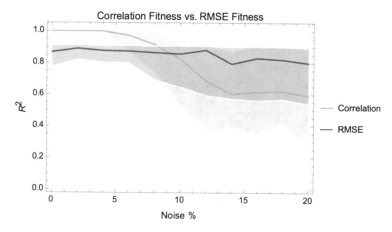

Fig. 8 Comparing using RMSE and correlation as the fitness function on Korns-8 as noise increases from 0 to 20% with 200 training points using R^2 as the metric for comparison

5.3   The Vladislavleva-1 Benchmark

Figure 9 compares the performance of using correlation and RMSE as fitness functions on the Vladislavleva-1 benchmark problem, ranging from 20 to 200 training data points with no noise. The correlation approach shows that it is able to consistently find models with better performance than when using RMSE as the fitness function.

Figure 10 explores how the two approaches vary in their sensitivity to noise. Both methods were given 200 training points and noise was varied from 0 to 20%. Correlation was observed to outperform RMSE as a fitness function until the noise level exceeded 6%. Beyond 6% the distributions of both methods widened significantly and the average performance of correlation as a fitness function became worse than RMSE at higher noise levels.

5.4   The Nguyen-5 Benchmark

The performance of correlation and RMSE as fitness functions was also compared using the Nguyen-5 benchmark problem. Figure 11 shows how they compare when no noise is present when training on varying number of points from 20 to 200. Correlation was observed to outperform RMSE as a fitness function for all of the cases between 20 and 200 points.

The noise tolerance of the two methods was also compared using the Nguyen-5 benchmark problem. Noise was varied from 0 to 20% with 200 training points. The results are shown in Fig. 12. The method using correlation as a fitness function

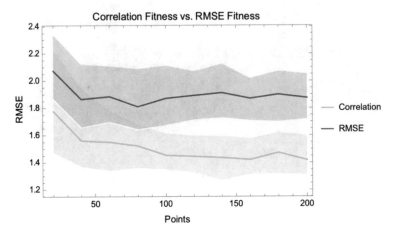

Fig. 9 Comparing using RMSE and correlation as the fitness function on Vladislavleva-1 with 0% noise as the number of points increases from 20 to 200

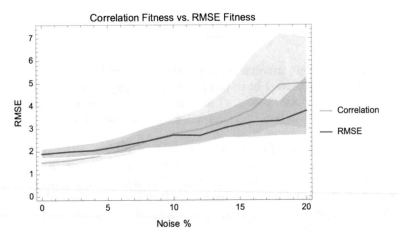

Fig. 10 Comparing using RMSE and correlation as the fitness function on Vladislavleva-1 as noise increases from 0 to 20% with 200 training points using RMSE as the metric for comparison

performed best until around 6% noise was present. Beyond 6% noise the two methods performed similarly with correlation having a slightly wider distribution of solutions.

5.5 Feynman Symbolic Regression Benchmark

The results of testing the performance of the two different fitness functions on the Feynman Symbolic Regression Benchmark are summarized in Table 2.

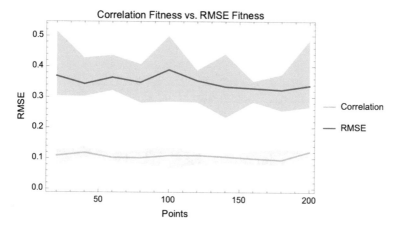

Fig. 11 Comparing using RMSE and correlation as the fitness function on Nguyen-5 with 0% noise as the number of points varies from 20 to 200

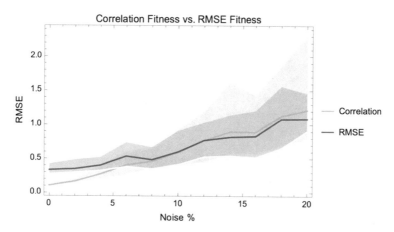

Fig. 12 Comparing using RMSE and correlation as the fitness function on Nguyen-5 as noise increases from 0 to 20% with 200 training points using RMSE as the metric for comparison

With just 3 data points and no noise, the correlation approach found better models in 38 of the 100 cases and tied in performance with the RMSE approach in 11 cases. A total of 21 problems were perfectly solved with just 3 data points using the correlation approach and 10 of those 21 were not perfectly solved using the RMSE approach with just 3 data points.

For example, Fig. 13 shows RMSE over generations on Eq. (22) (# 8 in [6]) when using correlation fitness.

$$\mu \times N_n. \tag{22}$$

Table 2 Feynman symbolic regression benchmark summary performance comparison of correlation against RMSE

Number of points	Noise %	Better	Tied	Perfectly solved	Perfectly solved where RMSE failed
3	0	38	11	21	10
3	10	27	0	0	0
20	0	82	17	41	24
20	10	78	0	0	0
200	0	81	17	46	29
200	10	79	0	0	0

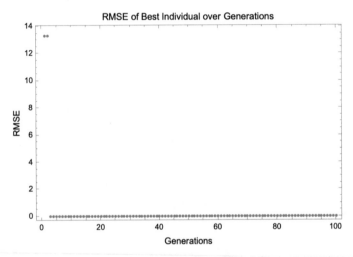

Fig. 13 The error of best individuals over generations is shown for Eq. (22) (# 8 in [6]) when using correlation as the fitness function. We can see that Eq. (22) is trivial and is solved almost immediately

RMSE converges to 0 very quickly in a sample evolutionary run, demonstrating that this equation is trivial to solve.

As another example, we take Eq. (23) (# 59 in [6]) as one where RMSE fitness does not converge to 0, see Fig. 14.

$$\frac{\varepsilon \times E^2}{2}. \tag{23}$$

As opposed to that, using the correlation fitness function results in runs like that depicted in Fig. 15. As we can see from the equations, these are very simple functional relationships. We shall examine more complicated ones later, but for now will look at noise.

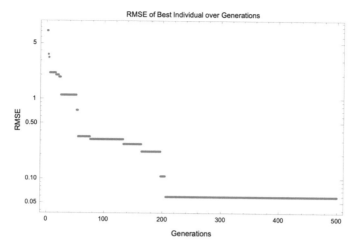

Fig. 14 The error of best individuals over generations is shown for Eq. (23) (# 59 in [6]) when using RMSE as the fitness function. We can see that progress seems to get stuck at a local minima and stops progressing around generation 200

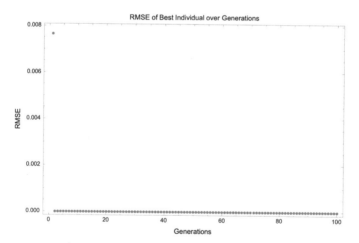

Fig. 15 The error of best individuals over generations is shown for Eq. (23) (# 59 in [6]) when using correlation as the fitness function. We can see that this problem now becomes trivial and is solved by the second generation when using correlation instead of RMSE as the fitness function

With just 3 data points and 10% noise added, the correlation approach does never converge to a perfect solution, but found better models in 27 of the 100 problems. On the other hand, RMSE is able to withstand the noise better and is able to converge to perfect solutions 13 times.

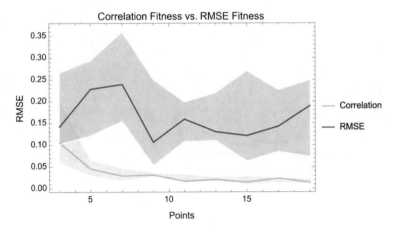

Fig. 16 Comparing using RMSE and correlation as the fitness function on Eq. (24) with 10% noise

Given that three data points really is an absolute minimum, it is no wonder that reviewers reacted incredulous to our results. But this seems more to be a reflection of the benchmark datasets, rather than the approach to solve it. In many cases, RMSE does also solve the problem perfectly with 3 data points (see Tables 4 and 5 in the appendix).

So while on trivial cases there might not be a difference between the performance of RMSE and correlation as a fitness function, in other cases, there is a difference, often a substantial one. With 20 points and no noise the correlation approach found better models in 82 of the 100 cases and was tied in another 17 cases. This means that the correlation approach beat or matched the performance of RMSE in 99 of the 100 cases of the Feynman benchmarks. As well, the correlation approach perfectly solved 41 of the problems. Of those 41, 24 were not solved using the RMSE approach.

With 10% noise added to the 20 data points, the correlation approach found better models in 78 of 100 cases. Here again, RMSE is better able to withstand the noise and converge to a perfect solution in 17 cases.

Finally, with 200 points and no noise, the correlation approach found better models than the RMSE approach in 81 of the 100 cases and was tied in 17 cases. This totals to 98 cases where the correlation approach either beat or matched the RMSE approach. A total of 46 cases were solved perfectly using the correlation approach. Of those 46 cases, 29 of them were not perfectly solved using the RMSE approach.

With 200 points and 10% noise added, the correlation approach found better models in 79 of the cases.

Figures 16, 17, and 18 show three examples of noisy data and the behavior of fitness functions depending on number of data points. These are non-trivial cases not perfectly solved by either of the methods even with 200 points. However, we can clearly see the progress by correlation as opposed to RMSE.

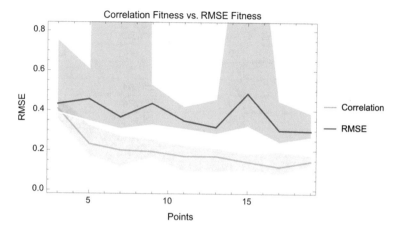

Fig. 17 Comparing using RMSE and correlation as the fitness function on Eq. (25) with 10% noise

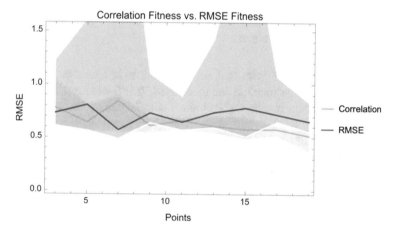

Fig. 18 Comparing using RMSE and correlation as the fitness function on Eq. (26) with 10% noise

The three equations used are

$$\frac{e^{\frac{-\theta^2}{2}}}{\sqrt{2\pi}} \tag{24}$$

$$\frac{e^{\frac{-(\theta/\sigma)^2}{2}}}{\sqrt{2\pi}\,\sigma} \tag{25}$$

Table 3 Sensitivity to constants in (22). Feynman Equation # 8

Constant	Number of Points	Noise %	RMSE	Correlation
1	20	10	0	1.35
1	20	0	0	0
5	20	10	261.69	8.74
5	20	0	276.76	0
50	20	10	3520.43	68.07
50	20	0	2151.56	0

and

$$\frac{e^{\frac{-((\theta-\theta_1)/\sigma)^2}{2}}}{\sqrt{2\pi}\sigma}. \tag{26}$$

5.6 Sensitivity to Constants

It was observed that Eq. (22) with 10% noise was able to be found with 0 error when using RMSE as the fitness function compared to an RMSE value of 1.35 when using correlation as the fitness function when trained with 20 points. The R^2 values from both approaches were 1, which indicates that they both found the correct general pattern in the data, but the scaling or position was off as a result of the alignment when using correlation as fitness function. To see if the problem difficulty was altered by introduction of a constant, Eq. (22) from the Feynman symbolic regression benchmark was modified by introducing a constant initially valued at 5 and then 50. The results are shown in Table 3. With no noise, the problem becomes much more difficult for the RMSE fitness function as the constant gets larger, while the problem difficulty does not change relative to the constant size when using correlation as the fitness function (recall Keijzer's observation mentioned earlier). When multiplicative noise is added, the performance when using correlation does get worse as the size of the constant increases, yet it strongly outperforms when compared to using RMSE as fitness function.

6 Discussion

Symbolic regression has been studied since Genetic Programming was invented more than 30 years ago. Numerous refinements have been proposed and examined, new benchmark datasets have been subjected to algorithmic variants, and results have been derived and considered in multiple forms. However, no measure has proven so efficient than simply replacing the fitness function based on absolute function values with a fitness measure that considers relative distance and leaves the absolute alignment of a functional model to a linear regression step in post-processing. Equipped with such a fitness function, many of the benchmark problems used (among them about half of the AI Feynman problems) can be safely removed from consideration as too easy. If 3 data points are sufficient to deduce the functional form of a model, then this is not a problem worthy of much attention.

For other problems, correlation did widely outperform RMSE as a fitness function, and should be the function of first choice in all regression problems. It remains to be seen whether there are some other factors that allow the algorithm discussed here to shine at solving the benchmark problems considered. For example, it could be that StackGP is particularly well suited for the task and the same might not be said of other GP systems.

7 Conclusions

What is clear is that the fitness function in any evolutionary algorithm has an extremely important role to play as far as the performance of the algorithm is concerned. Lessons derived from the success of correlation-based fitness in symbolic regression might transfer to other tasks, like classification problems. At the very least, it could facilitate the solution of harder problems in other domains, by training a researcher's eye on the essential functions of a system that need to be tuned to arrive at feasible solutions.

Acknowledgements Part of this work was funded by the Koza Endowment to MSU. Computer support by MSU's iCER high-performance computing center is gratefully acknowledged.

8 Appendix

This appendix lists all 100 AI Feynman problems and their solution using correlation and RMSE as fitness functions, for 0 and 10% noise levels at 3 data points only (Tables 4 and 5). More performance results are available online at https://tinyurl.com/stackGPGPTP.

Table 4 The resulting RMSE values on test data when using correlation or RMSE as the fitness function during training for first 50 Feynman equations. Three random training data points were used with either 0 or 10% noise

Filename	EQ #	Correlation 3 Pts 0%	RMSE 3 Pts 0%	Correlation 3 Pts 10%	RMSE 3 Pts 10%
I.6.20a	1.	**0.032**	0.304	**0.12**	0.35
I.6.20	2.	**0.41**	0.53	**0.41**	0.43
I.6.20b	3.	**0.68**	0.83	**0.57**	0.61
I.8.14	4.	21.71	**21.61**	**11.2**	15.66
I.9.18	5.	**2.02**	2.09	**2.15**	2.24
I.10.7	6.	2.29	**1.63**	2.74	**1.95**
I.11.19	7.	155.4	**137.08**	145.9	**142.88**
I.12.1	8.	0.	0.	3.74	**0.**
I.12.2	9.	291.77	**235.69**	302.15	**244.97**
I.12.4	10.	1.29	**1.17**	1.25	**0.95**
I.12.5	11.	0.29	**0.26**	0.29	**0.26**
I.12.11	12.	0.	0.	4.48	**0.**
I.13.4	13.	**311.99**	382.06	380.01	**378.98**
I.13.12	14.	123.88	**123.**	131.49	**121.92**
I.14.3	15.	0.	0.	16.67	**0.**
I.14.4	16.	**0.**	29.75	**9.03**	29.53
I.15.3x	17.	9.06	**6.44**	9.83	**7.16**
I.15.3t	18.	2.15	**1.81**	2.31	**2.1**
I.15.1	19.	4.34	**3.57**	4.91	**3.54**
I.16.6	20.	**7.26**	8.16	**8.41**	9.61
I.18.4	21.	**5.2**	6.38	**5.73**	6.4
I.18.12	22.	**39.64**	81.53	88.04	**0.**
I.18.14	23.	289.04	**269.43**	312.51	**294.09**
I.24.6	24.	186.77	**103.88**	172.39	**111.05**
I.25.13	25.	0.	0.	0.61	**0.**
I.26.2	26.	1.44	**0.**	4.78	**0.**
I.27.6	27.	4.59	**3.53**	**3.47**	3.82
I.29.4	28.	0.	0.	0.88	**0.**
I.29.16	29.	42.96	**39.75**	44.78	**39.01**
I.30.3	30.	170.3	**47.**	56.17	**45.07**
I.30.5	31.	0.	0.	1.07	**0.**
I.32.5	32.	**10.48**	11.09	11.02	**10.95**
I.32.17	33.	**58.75**	62.13	74.19	**63.58**
I.34.8	34.	118.78	**117.31**	152.25	**112.47**
I.34.1	35.	12.5	**12.39**	12.833	**12.611**
I.34.14	36.	9.69	**6.41**	9.27	**8.09**
I.34.27	37.	0.	0.	4.64	**0.**
I.37.4	38.	**45.37**	53.82	**34.71**	44.55
I.38.12	39.	809.61	**764.83**	797.34	**766.62**
I.39.1	40.	**0.**	33.53	**6.12**	49.26
I.39.11	41.	40.23	**22.69**	32.95	**29.32**
I.39.22	42.	**57.45**	109.84	136.6	**116.42**
I.40.1	43.	3.21×10^{12}	3.12×10^{12}	2.3×10^{15}	9.67×10^{14}
I.41.16	44.	31.63	**29.5**	31.62	**30.15**
I.43.16	45.	**93.43**	123.12	140.63	**113.85**
I.43.31	46.	0.	0.	15.21	**0.**
I.43.43	47.	24.15	**20.21**	23.18	**19.87**
I.44.4	48.	315.51	**278.57**	299.86	**265.23**
I.47.23	49.	**5.43**	6.32	7.34	**4.88**
I.48.2	50.	30.91	**17.23**	136.99	**29.78**

Table 5 The resulting RMSE values on test data when using correlation or RMSE as the fitness function during training for the second set of 50 Feynman equations. Three random training data points were used with either 0 or 10% noise

Filename	EQ #	Correlation 3 Pts 0%	RMSE 3 Pts 0%	Correlation 3 Pts 10%	RMSE 3 Pts 10%
I.50.26	51.	**32.39**	32.81	35.81	**33.07**
II.2.42	52.	96.89	**67.68**	111.34	**93.28**
II.3.24	53.	**0.**	0.22	**0.044**	0.328
II.4.23	54.	**0.**	0.29	**0.339**	0.357
II.6.11	55.	**0.27**	0.28	**0.279**	0.362
II.6.15a	56.	5.59	**4.05**	5.01	**4.11**
II.6.15b	57.	40.16	**39.29**	29.19	**28.66**
II.8.7	58.	**1.06**	1.24	**1.106**	1.116
II.8.31	59.	**0.**	35.63	**9.68**	23.01
II.10.9	60.	2.97	**1.52**	2.74	**1.79**
II.11.3	61.	**1.135**	1.155	1.25	**1.08**
II.11.17	62.	27.32	**24.32**	**20.98**	22.98
II.11.20	63.	118.25	**94.68**	103.66	**95.24**
II.11.27	64.	5.24	**2.61**	4.21	**2.71**
II.11.28	65.	**0.012**	0.2	0.77	**0.52**
II.13.17	66.	**0.3047**	0.3393	0.36	**0.3**
II.13.23	67.	2.12	**1.59**	2.68	**1.93**
II.13.34	68.	3.9	**2.76**	5.75	**3.76**
II.15.4	69.	**67.4**	88.35	95.53	**83.07**
II.15.5	70.	**63.28**	80.92	94.52	**92.21**
II.21.32	71.	0.714	**0.656**	0.67	**0.66**
II.24.17	72.	**1.72**	2.31	**2.405**	2.963
II.27.16	73.	0.	0.	79.1	**0.**
II.27.18	74.	0.	0.	17.51	**0.**
II.34.2a	75.	**0.**	5.86	**4.33**	5.04
II.34.2	76.	**0.**	47.1	**9.4**	45.
II.34.11	77.	65.58	**62.45**	73.52	**63.65**
II.34.29a	78.	**0.**	3.07	**1.98**	2.996
II.34.29b	79.	569.68	**454.52**	534.74	**460.95**
II.35.18	80.	**4.73**	5.1	**4.99**	5.36
II.35.21	81.	65.37	**49.**	58.39	**48.6**
II.36.38	82.	18.52	**16.24**	18.7	**15.76**
II.37.1	83.	136.845	**46.77**	92.81	**45.08**
II.38.3	84.	122.17	**114.56**	148.47	**98.47**
II.38.14	85.	**0.76**	0.99	**0.88**	1.09
III.4.32	86.	11.23	**7.24**	9.05	**8.04**
III.4.33	87.	41.59	**27.14**	42.31	**33.12**
III.7.38	88.	**0.**	50.43	55.54	**54.95**
III.8.54	89.	**3.68**	4.38	**4.1**	4.6
III.9.52	90.	45.78	**38.94**	46.29	**39.17**
III.10.19	91.	69.8	**54.84**	72.05	**56.28**
III.12.43	92.	0.	0.	4.08	**0.**
III.13.18	93.	1152.36	**1107.82**	1206.76	**1089.88**
III.14.14	94.	109.04	**98.87**	110.38	**97.74**
III.15.12	95.	105.83	**105.45**	107.98	109.96
III.15.14	96.	8.06	**7.1**	7.84	**7.08**
III.15.27	97.	**0.**	33.66	**28.29**	35.75
III.17.37	98.	88.19	**72.66**	90.	**74.54**
III.19.51	99.	**0.61**	0.72	**0.59**	0.86
III.21.20	100.	**126.007**	131.52	144.83	**132.95**

References

1. Koza, J.R.: Genetic Programming: On the Programming of Computers By Means Of Natural Selection. MIT Press, Cambridge, MA (1992)
2. Banzhaf, Wolfgang, Nordin, Peter, Keller, Robert E., Francone, Frank: Genetic Programming-An Introduction. Morgan Kaufmann Publishers, San Francisco, CA (1998)
3. White, D.R., Arcuri, A., Clark, J.A.: Evolutionary improvement of programs. IEEE Trans. Evol. Comput. **15**(4), 515–538 (2011)
4. Petke, J., Haraldsson, S.O., Harman, M., Langdon, W.B., White, D.R., Woodward, J.R.: Genetic improvement of software: a comprehensive survey. IEEE Trans. Evol. Comput. **22**(3), 415–432 (2017)
5. Affenzeller, M., Winkler, S., Wagner, S., Beham, A.: Genetic Algorithms and Genetic Programming-Modern Concepts and Practical Applications. CRC Press/Taylor & Francis, New York, NY (2009)
6. Udrescu, S.-M., Tegmark, M.: A physics-inspired method for symbolic regression. Sci. Adv. **6**, eaay2631 (2020)
7. Livadiotis, G., McComas, D.J.: Fitting method based on correlation maximization: applications in space physics. J. Geophys. Res. Space Phys. **118**, 2863–2875 (2013)
8. Livadiotis, G.: General fitting methods based on l_q norms and their optimization. Stats **3**, 16–31 (2020)
9. Keijzer, M.: Improving symbolic regression with interval arithmetic and linear scaling. In: European Conference on Genetic Programming, pp. 70–82. Springer (2003)
10. Nicolau, M., McDermott, J.: Genetic programming symbolic regression: What is the prior on the prediction? In: Banzhaf, W., Goodman, E., Sheneman, L., Trujillo, L., Worzel, B. (eds.) Genetic Programming Theory and Practice XVII, pp. 201–225. Springer (2020)
11. White, D.R., Mcdermott, J., Castelli, M., Manzoni, L., Goldman, B.W., Kronberger, G., Jaśkowski, W., O'Reilly, U.-M., Luke, S.: Better gp benchmarks: community survey results and proposals. Genet. Program. Evol. Mach. **14**(1), 3–29 (2013)
12. Korns, M.F.: A baseline symbolic regression algorithm. In: Genetic Programming Theory and Practice X, pp. 117–137. Springer (2013)
13. Korns, M.F.: Extreme accuracy in symbolic regression. In: Genetic Programming Theory and Practice XI, pp. 1–30. Springer (2014)
14. Korns, M.F.: Extremely accurate symbolic regression for large feature problems. In: Genetic Programming Theory and Practice XII, pp. 109–131. Springer (2015)
15. Vladislavleva, E.J., Smits, G.F., Hertog, D.D.: Order of nonlinearity as a complexity measure for models generated by symbolic regression via pareto genetic programming. IEEE Trans. Evol. Comput. **13**(2), 333–349 (2008)
16. Uy, N.Q., Hoai, N.X., O'Neill, M., McKay, R.I., Galván-López, E.: Semantically-based crossover in genetic programming: application to real-valued symbolic regression. Genet. Program. Evol. Mach. **12**(2), 91–119 (2011)
17. McDermott, J., White, D.R., Luke, S., Manzoni, L., Castelli, M., Vanneschi, L., Jaskowski, W., Krawiec, K., Harper, R., De Jong, K., et al.: Genetic programming needs better benchmarks. In: Proceedings of the 14th Annual Conference on Genetic and Evolutionary Computation, pp. 791–798 (2012)
18. Korns, M.F.: Large-scale, time-constrained symbolic regression. In: Genetic Programming Theory and Practice IV, pp. 299–314. Springer (2007)
19. Korns, M.F.: Large-scale, time-constrained symbolic regression-classification. In: Genetic Programming Theory and Practice V, pp. 53–68. Springer (2008)
20. Korns, M.F., Nunez, L.: Profiling symbolic regression-classification. In: Genetic Programming Theory and Practice VI, pp. 1–14. Springer (2009)
21. Korns, M.F.: Symbolic regression of conditional target expressions. In: Genetic Programming Theory and Practice VII, pp. 211–228. Springer (2010)
22. Korns, M.F.: Abstract expression grammar symbolic regression. In: Genetic Programming Theory and Practice VIII, pp. 109–128. Springer (2011)

23. Feynman, R., Leighton, R., Sands, M.: The Feynman Lectures on Physics, vol. 1. Basic Books, New York, NY (1963)
24. Feynman, R., Leighton, R., Sands, M.: The Feynman Lectures on Physics, vol. 2. Addison Wesley, Boston, MA (1963)
25. Feynman, R., Leighton, R., Sands, M.: The Feynman Lectures on Physics, vol. 3. Addison Wesley, Boston, MA (1963)

GUI-Based, Efficient Genetic Programming and AI Planning for Unity3D

Robert Gold, Andrew Haydn Grant, Erik Hemberg, Chathika Gunaratne, and Una-May O'Reilly

Abstract We present a GUI-driven and efficient Genetic Programming (GP) and AI Planning framework designed for agent-based learning research. Our framework, ABL-Unity3D, is built in Unity3D, a game development environment. ABL-Unity3D addresses challenges entailed in co-opting Unity3D: making the simulator serve agent learning rather than humans playing a game, lowering fitness evaluation time to make learning computationally feasible, and interfacing GP with an AI Planner to support hybrid algorithms. We achieve this by developing a Graphical User Interface (GUI) using the Unity3D editor's programmable interface and performance optimizations. These optimizations result in at least a 3x speedup. In addition, we describe ABL-Unity3D by explaining how to use it for an example experiment using GP and AI Planning. We benchmark ABL-Unity3D by measuring the performance and speed of the AI Planner alone, GP alone, and the AI Planner with GP.

R. Gold (✉) · A. H. Grant · E. Hemberg · C. Gunaratne · U.-M. O'Reilly
ALFA, MIT CSAIL, Cambridge, MA, USA
e-mail: robertgold@csail.mit.edu

A. H. Grant
e-mail: haydn@mit.edu

E. Hemberg
e-mail: hembergerik@csail.mit.edu

C. Gunaratne
e-mail: contact@chathika.com

U.-M. O'Reilly
e-mail: unamay@csail.mit.edu

© The Author(s), under exclusive license to Springer Nature Singapore Pte Ltd. 2023
L. Trujillo et al. (eds.), *Genetic Programming Theory and Practice XIX*,
Genetic and Evolutionary Computation, https://doi.org/10.1007/978-981-19-8460-0_3

57

1 Introduction

Many computational simulation environments support the development and evaluation of agent-based learning (ABL) [1] algorithms, both academic and commercial
[2]. While many projects center on the learning of gameplay, there has been growing
interest in agent-based learning in non-game contexts, such as network security and
attack planning or Covid-19 contagion interaction protocols [3, 4]. This implies the
need to simulate more realistic environments. For this, well-supported and capable
game development environments exist, such as Unity3D [5]. However, these often
center on human gameplay, not AI agent-based learning. Further, game development
environments co-opted to non-game playing agents present the risk of expensive
fitness evaluation and inappropriate interfaces for developing and evaluating agent-
based learning algorithms. There has been some progress toward integrating AI
agent-based learning in Unity, such as ML-Agents which is an agent-based learning
reinforcement learning library [6].

We provide a Genetic Programming (GP) [7] and AI Planning [8] framework for
Unity3D. We introduce ABL-Unity3D in an attempt to reduce development time
and effort and increase performance. Our primary goal is to provide an open-source,
domain-targetable agent-based learning system that provides tools for GP and AI
Planning research and development, addresses agent evaluation efficiency, and pivots
the human interaction lens of Unity3D to an AI-developer user-friendly GUI. We
want the system to support AI experiments that are easy to design and efficient to
execute. We aim to introduce a baseline for running AI experiments that require 3D
simulation, not a unifying framework for all possible AI experiments. In this paper,
we present ABL-Unity3D, which is our project centered on these goals. The code
repository for ABL-Unity3D can be found at [9].

Our main contribution is a simulator that integrates directly with GP and AI Planning, is easy to configure, has transparent AI components, and is domain-targetable.
In addition, we examine the ability of GP and AI Planning in ABL-Unity3D to find
high-quality solutions and benchmark the speed of ABL-Unity3D with and without
optimizations.

The paper is structured as follows. In Sect. 2 we describe ABL-Unity3D. In Sect. 3,
we specify and simulate an example scenario as a use case for ABL-Unity3D. In
Sect. 4 we discuss our findings. Finally, in Sect. 5, we provide a summary and future
work.

2 ABL-Unity3D

ABL-Unity3D is designed to be efficient and user-friendly. For example, conducting
a GP experiment requires no coding. It only requires setting parameters in GUI
fields, such as selecting GP primitives and a fitness function, though creating new

GP primitives and fitness functions requires coding. The primary design objectives for ABL-Unity3D are

Efficiency via caching, fast copying, and multithreading.
Extendability via object-oriented design and decoupling the AI and simulation components.
Usability via a GUI.

In this section, we give an overview of ABL-Unity3D. In Fig. 1, we present an overview of the components in ABL-Unity3D. The components are as follows.

Simulator (Sect. 2.1) ABL-Unity3D interfaces Unity3D with Genetic Programming and AI Planning to support specific methodological investigations in agent-based learning. ABL-Unity3D uses the Unity3D game engine as a simulator and, thus, is written in C#. Unity3D provides an easy-to-use, efficient, and extendable GUI and API interface to run simulations and AI.

AI (Sects. 2.2 and 2.3) ABL-Unity3D contains a simulation world state. This state can be input to the GP and AI Planner. The GP and AI Planner can, but are not required to, act as inputs to (or outputs from) each other to create a hybrid algorithm. The outputs of GP (typically candidate solutions that need evaluation in a 3D simulator) and the outputs of the AI Planner (typically plans for agents to follow in a 3D simulator) become inputs to the simulation. As a simulation progresses, the world state updates. The AI then receives these updates.

GUI (Sect. 2.4) ABL-Unity3D provides a GUI to set parameters for and examine the simulator and learning components. The ABL-Unity3D GUI uses the Unity3D editor to provide the ability to view the simulation as it is running and modify it. The GUI also makes it easy to run, organize and design repeatable experiments, and save results.

Finally, in Sect. 2.5 we describe the optimizations we have done for improving the efficiency of ABL-Unity3D.

2.1 Simulator

ABL-Unity3D uses the Unity3D game engine. It is widely used [10] and provides a large and well-documented API [11] for designing simulations and GUIs. Unity3D has a large community-driven base that provides third-party functionality, which is easily integrated using the Unity Asset Store [12]. The main simulator concepts are

Simulation World State The simulator in ABL-Unity3D maintains and gives access to a world state that describes the properties of the simulation. The simulation world state can contain two types of objects: agents and agent actions. There are two types of agents: SimAgent and SimGroup. SimAgent is a single agent within the world state, and a SimGroup is a group of SimAgents. The other type of objects, agent actions, are called SimAction.

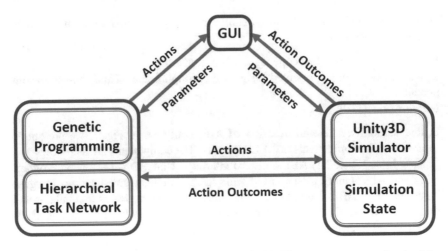

Fig. 1 An overview of the components in ABL-Unity3D. Key components are the simulator (Unity3D), AI (GP and AI planning), and GUI

Terrain and Pathfinding ABL-Unity3D allows the user to design a 3D terrain for agents to traverse. ABL-Unity3D uses a third-party A* agent pathfinding library to allow agents to traverse terrain [13].

Parameters ABL-Unity3D provides parameters for the user to configure, which define how a simulation should run. We describe one parameter which is relevant to the example experiment in Sect. 3. `secondsPerSimStep` defines the number of seconds to simulate per step of the simulation. The lower the `secondsPerSimStep` is, the greater is the granularity of the simulation. In turn, this also means the simulation will run at a slower pace.

2.2   Genetic Programming

One key innovation of GP in ABL-Unity3D is the GP solution and fitness function implementations which are designed to achieve a fast, extendable, and user-friendly GP framework. Another key innovation is the ability of the GP to interface with the AI Planner, Simulator, and GUI.

2.2.1   GP Representations

ABL-Unity3D provides a standard strongly typed GP implementation with the option for different initialization methods and genetic operators. ABL-Unity3D represents GP solutions as tree data structures [14]. Each GP node in the tree has an evaluation type. When ABL-Unity3D evaluates a GP node, it returns that type. Figure 2 shows

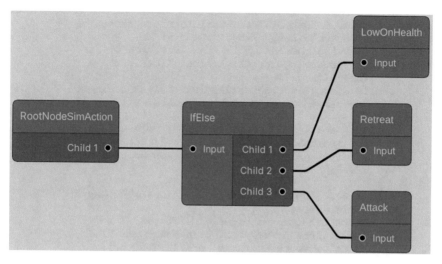

Fig. 2 Visualization of a GP solution that represents a conditional statement which evaluates to a SimAction in ABL-Unity3D

a visualization (generated by ABL-Unity3D) of a GP primitive written to represent a conditional statement. This implementation of a conditional statement evaluates to a SimAction.

The root node of the tree represents the evaluation type of the tree. The root node must have one child node with the same evaluation type as the root node. For example, the conditional statement in Fig. 2 evaluates to a SimAction. The root node type "RootNodeSimAction" shows this. The other nodes can have any number of child nodes and be named anything.

GpBuildingBlock<T> (line 1, Fig. 3) is the super-class for GP primitives where T is the evaluation type. GpBuildingBlock<T>.children is the list of child nodes. Child nodes extend GpBuildingBlock<T>. A GP primitive must define a constructor which passes all child nodes to the primitive base constructor. A GP primitive must also override the function Evaluate, which returns the result of the GP primitive upon evaluation. The return type of Evaluate must be the same as the evaluation type T for the GP primitive.

Figure 3 shows sample code for defining the conditional statement GP primitive shown in Fig. 2. The GP primitive Conditional has an evaluation type of SimAction (line 1, Fig. 3). Lines 2–4 are helper properties used for visualization. Lines 6–9 define the constructor for Conditional class. Lastly, lines 11–12 define the Evaluate method. Line 12 first evaluates the child node, which represents the condition for the conditional. If true, it returns the evaluation of the child node TrueBranch. Otherwise, it returns the evaluation of the child node FalseBranch.

```
1   public class Conditional : GpBuildingBlock<SimAction> {
2       public GpBuildingBlock<bool> Cond =>
            (GpBuildingBlock<bool>) this.children[0];
3       public GpBuildingBlock<SimAction> TrueBranch =>
            (GpBuildingBlock<SimAction>) this.children[1];
4       public GpBuildingBlock<SimAction> FalseBranch =>
            (GpBuildingBlock<SimAction>) this.children[2];
5
6       public Conditional(GpBuildingBlock<bool> cond,
            GpBuildingBlock<SimAction> trueBranch,
            GpBuildingBlock<SimAction> falseBranch) : base(cond,
            trueBranch, falseBranch) { }
7
8       public override SimAction Evaluate(GpFieldsWrapper
            gpFieldsWrapper) {
9         return Cond.Evaluate(gpFieldsWrapper) ?
              TrueBranch.Evaluate(gpFieldsWrapper) :
              FalseBranch.Evaluate(gpFieldsWrapper);
10      }
11  }
```

Fig. 3 Sample code to define a GP primitive for a conditional statement which evaluates to a
SimAction

Child nodes of a GP primitive can be immutable; genetic operators can not modify
them. For example, replace line 5 in Fig. 3 with a hard-coded instance of a subclass
of GpBuildingBlock<T>, ex. the GP primitive "Attack".

2.2.2 Fitness Function

To define a fitness function, define a class that extends FitnessFunction. In
addition, the sub-class must implement several different interfaces. It must implement IAsync or ISync depending on whether the fitness function is asynchronous.
It must also implement IUsesSuppliedSimWorldState or ICreatesSim
WorldState, which defines whether the fitness function takes the simulation
world state as an input or whether it generates its own. Lastly, if the fitness function works with the AI Planner, it must implement the interface IUsesPlanner.
Note that the sub-class must implement a constructor, and the evaluation function
EvaluateIndividual/EvaluateIndividualAsync.

2.2.3 Parameters

ABL-Unity3D provides the basic parameters for strongly typed GP. In addition, ABL-
Unity3D provides the ability to filter for unique individuals in initialization through
the parameter onlyUnique.

2.3 AI Planning

ABL-Unity3D implements a hierarchical task network (HTN) [15] that generates all possible plans for agent behavior instead of finding a single plan that satisfies the given goal method. To do this, the planner decomposes a goal method into a set of sub-methods that achieve the given goal. The methods chosen are then implemented using concrete `SimActions`, which then execute in the simulation. It is possible to use the AI Planner with GP in ABL-Unity3D. The AI Planner generates plans depth-first, but this can be modified.

For example, suppose we have the "Attack Area" action, which sends agents to attack enemy agents in a specified area. This action could decompose into the sequence of actions "Approach Area" and then "Engage Enemy". The action "Approach Area" could then decompose into the action "Approach in Single Group" or the action "Approach in Multiple Groups". The HTN will look at both possibilities and choose the one that performs the best. See Fig. 4 for a visualization. Note that we use the terms AI Planner and HTN interchangeably.

2.4 Graphical User Interface (GUI)

In order to be more user-friendly, the ABL-Unity3D GUI attempts to facilitate interaction with the simulator and AI; see Fig. 5. The ABL-Unity3D GUI capabilities include, but are not limited to (Figs. 6).

- Defining simulation parameters. See Fig. 9;
- Running GP experiments and AI Planning experiments. Figure 12 shows the UI for defining a GP experiment. Figure 8 shows the UI for defining the parameters to the AI Planner;
- Examination of results. Figure 10 shows the UI for examining the results of a GP experiment. Figure 7 shows the UI for examining the results of the AI Planner;
- Visualization of GP and AI Planner results. Figure 2 shows the UI for viewing a GP individual. The AI Planner uses the same UI;

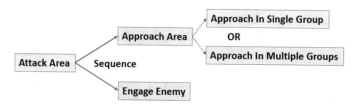

Fig. 4 A depiction of the possible decompositions for an action that sends agents to attack enemy agents in a specified area

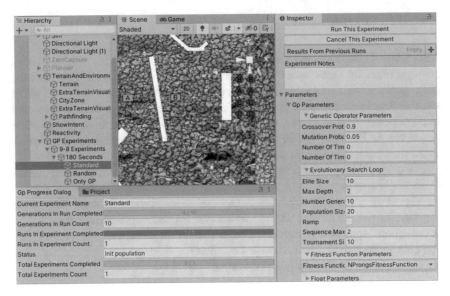

Fig. 5 A sample view of a simulation and GP GUI

- Examination of the simulation world state and results generated from experiments. Figure 11 shows the UI to examine agent attributes;
- Visualization of the actions an agent performs as the scenario simulates. See Fig. 6.

The ABL-Unity3D GUI implementation is object-oriented and exposes properties to the user. ABL-Unity3D makes it easy to swap out the simulation because the simulation world state is decoupled from the GP and the AI Planning. The only aspects of the GP and AI Planning components that depend on the simulation world state are user-defined GP primitives, fitness functions, and AI Planning methods (Fig. 12).

2.5 Optimizations

Generally, the simulation world state, and objects contained in it, are object-oriented. However, simulation speed is a priority. Thus, ABL-Unity3D implements an abstraction similar to Model-View-Controller (MVC) to improve efficiency. This abstraction allows for faster copying of the simulation world state. SimUnit is similar to the "model" in MVC, and SimUnitFollower is like the "view". We do not make a strict separation for the Controller part of MVC, so both classes implement controller-like behavior. Another optimization ABL-Unity3D implementing is multithreading. ABL-Unity3D runs AI Planning, GP, agent pathfinding, and the simulation in separate parallel processing threads allowing these processes to be run on different

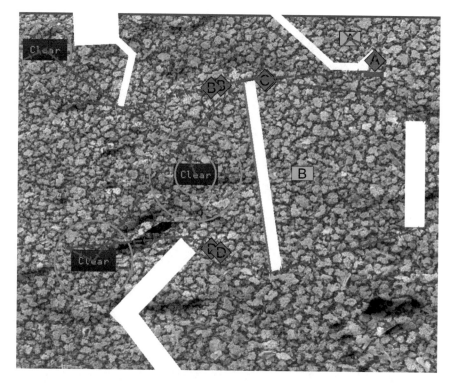

Fig. 6 The visualization of agents performing actions as a scenario is simulating

threads and canceled independently. Lastly, to improve performance for pathfinding, ABL-Unity3D caches previously queried paths from different locations in the terrain.

3 ABL-Unity3D Example Scenario: N Prongs

In this section, we evaluate the execution speed and performance of ABL-Unity3D. In Sect. 3.1, we describe an example scenario in ABL-Unity3D, which aims to generate strategic plans. We call this the *N* Prongs Scenario. In Sect. 3.2, we propose a solution specification and define the fitness evaluation for a solution. Section 3.3 provides benchmark performance of various search algorithms in ABL-Unity3D.

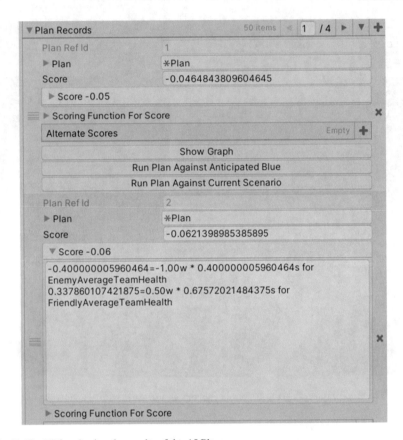

Fig. 7 The UI for viewing the results of the AI Planner

3.1 Scenario Specification

Consider the following scenario:

> "Blue" team is defending a location, and "Red" team is attempting to take over that location. For the "Red" team to take over the location that the "Blue" team is defending, the "Red" team must damage the "Blue" team units as much as possible. However, it is also crucial for at least some of the "Red" team units to survive.

In other words, the main goal is for the "Red" team to eliminate the "Blue" team and occupy the city, and a secondary goal is to minimize the damage done to agents on the "Red" team.

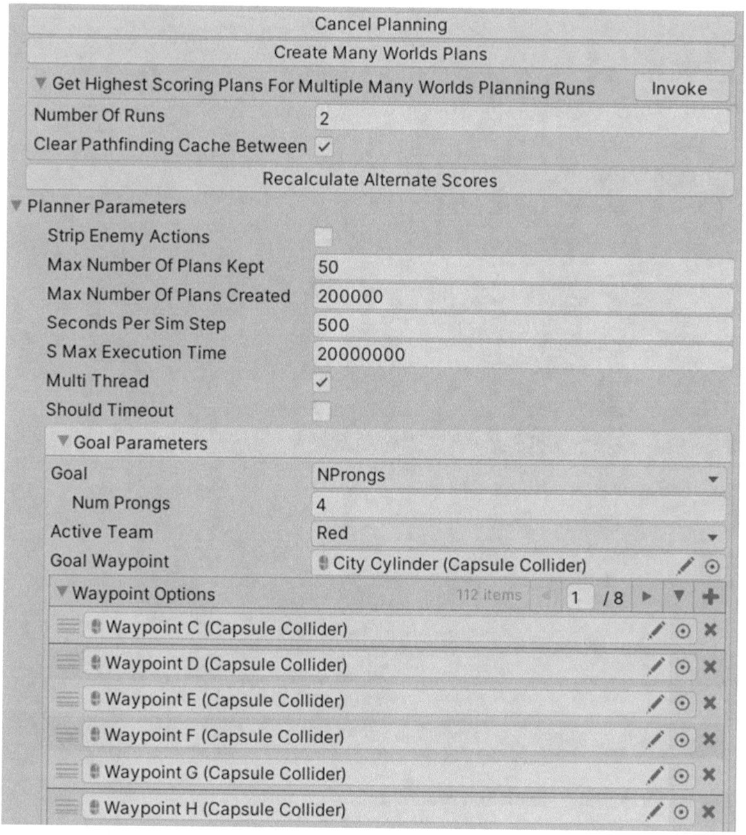

Cancel Planning	
Create Many Worlds Plans	
▼ Get Highest Scoring Plans For Multiple Many Worlds Planning Runs	Invoke
Number Of Runs	2
Clear Pathfinding Cache Between	✓
Recalculate Alternate Scores	
▼ Planner Parameters	
Strip Enemy Actions	☐
Max Number Of Plans Kept	50
Max Number Of Plans Created	200000
Seconds Per Sim Step	500
S Max Execution Time	20000000
Multi Thread	✓
Should Timeout	☐
▼ Goal Parameters	
Goal	NProngs ▾
Num Prongs	4
Active Team	Red ▾
Goal Waypoint	🢖 City Cylinder (Capsule Collider) ✏ ⊙
▼ Waypoint Options	112 items ◄ 1 /8 ► ▼ ✚
☰ 🢖 Waypoint C (Capsule Collider)	✏ ⊙ ✖
☰ 🢖 Waypoint D (Capsule Collider)	✏ ⊙ ✖
☰ 🢖 Waypoint E (Capsule Collider)	✏ ⊙ ✖
☰ 🢖 Waypoint F (Capsule Collider)	✏ ⊙ ✖
☰ 🢖 Waypoint G (Capsule Collider)	✏ ⊙ ✖
☰ 🢖 Waypoint H (Capsule Collider)	✏ ⊙ ✖

Fig. 8 The UI for defining parameters for the AI Planner

3.2 Solution Specification

We first propose a solution specification for this scenario and then define how the fitness of a solution is evaluated. The approach we use to solve this scenario is to split the "Red" team into N groups. These groups can then attack the location from N different waypoints. In particular, each group can travel through and attack one waypoint along its path to the location. We call these waypoints "prongs". We call this the N Prongs Scenario.

Let C be the prongs chosen, and P be the set of prongs to choose from, such that

$$P = \{\mathbf{x}_0, \ldots, \mathbf{x}_n\}, \mathbf{x}_i \in \mathbb{R}^2, \ C \subseteq P.$$

There must be greater than or equal to N prongs in P, $|P| \geq N$. There must be between one and N prongs in C, $1 \leq |C| \leq N$. Altogether, we can write $1 \leq |C| \leq N \leq |P|$.

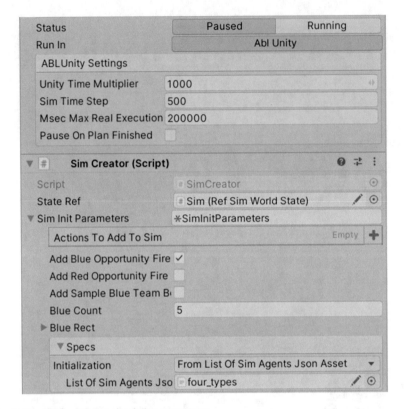

Fig. 9 The UI for defining simulation parameters

Let R be the set of "Red" team agents, B the set of "Blue" team agents, and $\mathscr{A}$ the set of all possible agents, such that $R = [r_0, \ldots, r_n]$, $B = [b_0, \ldots, b_m]$, where $r_i \in \mathscr{A}, b_i \in \mathscr{A}, R \cap B = \emptyset$.

We must assign a mapping of "Red" team agents to prongs. Let us call this mapping M.

$$M : \mathscr{A}^n \times \mathbb{R}^{k \times 2} \to (\mathscr{A}, \mathbb{R}^2)^n,$$

where $M(R, P) = [(a_0, \mathbf{x}_i), \ldots, (a_n, \mathbf{x}_j)]$, and $0 \leq i, \ j \leq |P| = k$.

Our objective is to maximize the total "Red" team health and minimize the total "Blue" team health. To achieve this, we must find the best prongs to choose and the best assignment of red team groups to those chosen prongs. Formally, our objective is to find an optimal C to maximize

$$f : (\mathscr{A}, \mathbb{R}^2)^n \times \mathscr{A}^m \to \mathbb{R},$$

where

$$f(M(R, P), B) = y$$

given R, P and B.

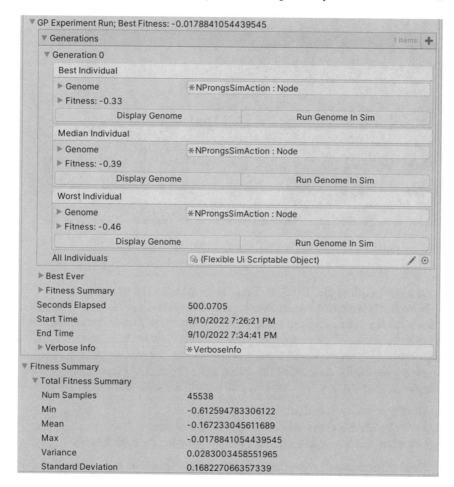

Fig. 10 The UI for viewing the results of a GP experiment

Fitness Evaluation We want to preserve "Red" team health while reducing "Blue" team health. In other words, if B is the set of blue team agents, and R is the set of red team agents, then our fitness function is

$$f(M(R, P), B) = 0.5 \cdot \overline{R}_{health} - \overline{B}_{health}$$

where

$$\overline{R}_{health} = 1/|R| \sum_{r \in R} r_{health}$$
$$\overline{B}_{health} = 1/|B| \sum_{b \in B} b_{health}$$

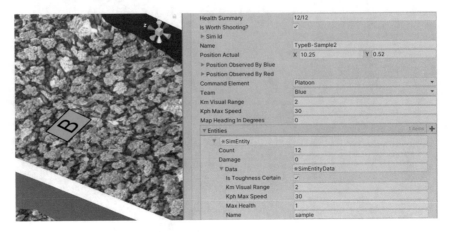

Health Summary	12/12	
Is Worth Shooting?	✓	
▶ Sim Id		
Name	TypeB-Sample2	
Position Actual	X 10.25	Y 0.52
▶ Position Observed By Blue		
▶ Position Observed By Red		
Command Element	Platoon	▾
Team	Blue	▾
Km Visual Range	2	
Kph Max Speed	30	
Map Heading In Degrees	0	
▾ Entities		1 items ✚
▾ ＊SimEntity		
Count	12	
Damage	0	
▾ Data	＊SimEntityData	
Is Toughness Certain	✓	
Km Visual Range	2	
Kph Max Speed	30	
Max Health	1	
Name	sample	

Fig. 11 The UI for viewing attributes of SimUnits, such as location and health, etc.

is the mean health of the "Red" or "Blue" team at the end of a simulation.

Search Algorithms Used in Benchmark Comparison ABL-Unity3D can generate solutions to this problem using GP, AI Planning, or a combination. Arguably, when $|P|$ is small, AI Planning is the most effective to solve this because it can feasibly exhaust the search space, though, if $|P|$ is large, the search space will increase in size exponentially. Exhausting the search space through AI Planning may take too long. To get around this, we look at search algorithms with time limits:

1. Use GP to derive C and M,
2. Use GP to derive C and AI Planning to derive M, and
3. Force the AI Planner to terminate before exhausting the search space.

None of these search algorithms are guaranteed to generate the optimal C and M, but it may be sufficient given time constraints.

3.3 Benchmark Performance

In this section, we benchmark the performance of ABL-Unity3D. We look at the following measurements:

1. The performance of different search algorithms in ABL-Unity3D, and
2. The speed of ABL-Unity3D.

All benchmarks are run on a PC with an i7 8700k CPU.

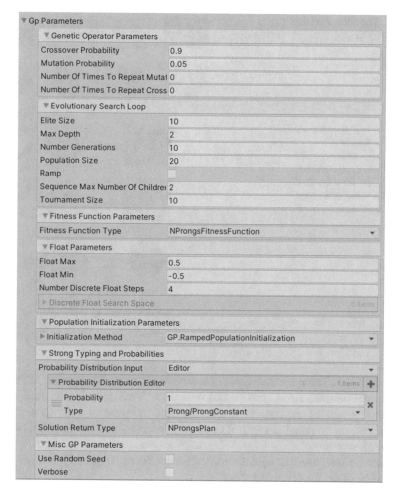

▼ Gp Parameters	
▼ Genetic Operator Parameters	
Crossover Probability	0.9
Mutation Probability	0.05
Number Of Times To Repeat Mutat	0
Number Of Times To Repeat Cross	0
▼ Evolutionary Search Loop	
Elite Size	10
Max Depth	2
Number Generations	10
Population Size	20
Ramp	☐
Sequence Max Number Of Childrer	2
Tournament Size	10
▼ Fitness Function Parameters	
Fitness Function Type	NProngsFitnessFunction
▼ Float Parameters	
Float Max	0.5
Float Min	-0.5
Number Discrete Float Steps	4
▶ Discrete Float Search Space	6 Items
▼ Population Initialization Parameters	
▶ Initialization Method	GP.RampedPopulationInitialization
▼ Strong Typing and Probabilities	
Probability Distribution Input	Editor
▼ Probability Distribution Editor	1 Items ✚
Probability	1
Type	Prong/ProngConstant
Solution Return Type	NProngsPlan
▼ Misc GP Parameters	
Use Random Seed	☐
Verbose	☐

Fig. 12 Part of the UI to define and run a GP experiment and view and save results

3.3.1 Solution Quality: Four Prongs Scenario

We benchmark the performance of ABL-Unity3D in finding high-quality solutions for the Four Prongs Scenario. We compare the following search algorithms:

- GP   — GP alone,
- Planner   — AI Planner alone,
- Planner + GP — AI Planner with GP, and
- Random   — Random search (as a baseline).

The key components for this benchmark are

Number of Prongs There are four prongs for the search algorithm to choose, $max(|C|) = 4$. Hence, the name Four Prongs.

Fig. 13 The different sprites for each agent type

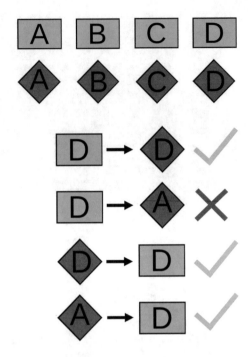

Fig. 14 "Red" team agents do the most damage to agents of the same type and very little to agents of other types. "Blue" team agents do an equal amount of damage to each agent type

Agent Types There are four types of agents: A, B, C, and D. See Fig. 13 for the sprites used for each agent type in ABL-Unity3D. "Red" team agents do the most damage to agents of the same type and very little to agents of other types. "Blue" team agents do an equal amount of damage to each agent type. See Fig. 14 for a visualization of this.

Agent Groups We split agents into four groups based on agent type. Each group attacks one of the four chosen prongs.

Agent Behaviors We hard-code "Blue" team behavior and evolve "Red" team behavior. The "Blue" team does not move and only attacks when a "Red" agent is within its weapon range. A "Red" agent's behavior is a sequence of ClearCircle actions. This action takes a set of agents and a circle on the map. The agents first move to this circle, then attack all agents within that circle.

Both "Red" and "Blue" team agents attack until the target agent loses all health.

Blue & Red Team Makeup The "Blue" team comprises five agents, and the "Red" team comprises eight. The "Blue" team comprises one agent of types B, C, and D, and two agents of type A. The red team comprises two agents of each type.

Terrain The terrain is a manually generated 3-dimensional environment with obstacles. The terrain and obstacles affect which paths agents can move along. In particular, agents must move around too steep terrain and all obstacles. ABL-Unity3D renders obstacles as white blocks. See Fig. 15 for a 3-dimensional rendering of the terrain, and Fig. 16 for a top-down 2-dimensional rendering.

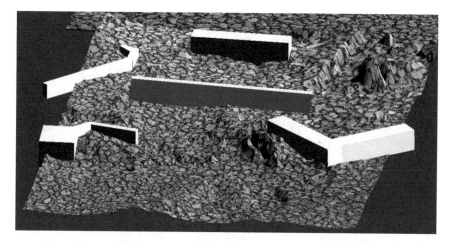

Fig. 15 A 3-dimensional rendering of the terrain used in this scenario. ABL-Unity3D renders obstacles as white blocks

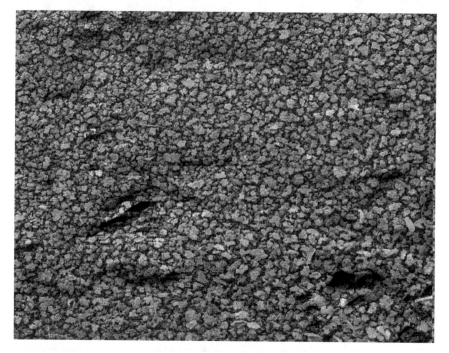

Fig. 16 A top-down 2-dimensional rendering of the terrain used in this scenario. ABL-Unity3D renders obstacles as white blocks

Fig. 17 The positions of agents on the "Blue" and "Red" team

Agent Locations The "Blue" team and "Red" team agents are located as shown in Fig. 17.

Waypoint Locations Figure 18 shows the location of each possible waypoint and the city. The green circles represent the waypoints, and the pink circle represents the city. There are 112 possible waypoints, $|P| = 112$, though only 22 are visible, 18 of which are empty, and four of which contain a single "Blue" team agent. The waypoints that are not visible are copies of those 18 waypoints which are empty. In essence, only 4 out of 112 waypoints contain "Blue" team agents.

Parameters We run each search algorithm for 600 ± 1 s. See Table 1 for the parameters used for each search algorithm.

Results We find that GP alone, on average, performs better than the other search algorithms. GP alone finds the highest performing individual with a fitness score of 0.27. We also see that the AI Planner with GP not only can perform better than the AI Planner alone but can also perform worse. The AI Planner always has the same score. In addition, Random Search can sometimes perform better than the AI Planner with GP and GP alone. Random search and the AI Planner with GP have the same mean, but the AI Planner with GP has a higher standard deviation (Table 2).

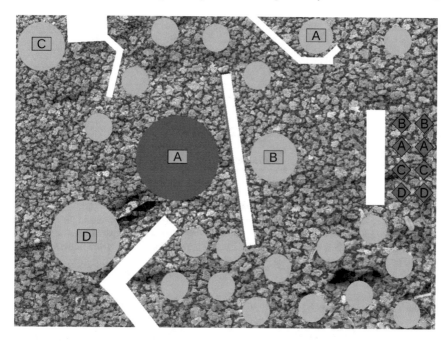

Fig. 18 The positions of each waypoint used in the Four Prongs experiment. The green circles represent each waypoint's location, shape, and size. The pink circle represents the location that the red team is trying to take over

Table 1 The parameters used to test the performance of each search algorithm in the Four Prongs Scenario

Parameter	GP	Planner + GP	Planner	Random
Elite size	50	10	N/A	N/A
Population size	200	30	N/A	N/A
Tournament size	50	10	N/A	N/A
Crossover probability	0.9	0.9	N/A	N/A
Mutation probability	0.05	0.05	N/A	N/A
Initialization	Only unique	Only unique	N/A	N/A
Seconds per sim step	500	500	500	N/A
Time limit (seconds)	600	600	600	600

Table 2 Solution score results for all four search algorithms for each run. Highest score is in **bold**

Run number	Random	Planner	Planner+GP	GP
1	−0.01	0.13	−0.16	**0.27**
2	−0.01	0.13	0.00	0.00
3	−0.02	0.13	0.15	0.25
4	0.12	0.13	0.16	0.13
5	−0.02	0.13	0.00	0.25
6	0.09	0.13	0.00	0.11
Mean±STD	0.03±0.06	0.13±0	0.03±0.11	0.15±0.09

Table 3 Average speed (seconds) and speedup using various optimizations (pathfinding caching and multithreading). We look at these measurements for the AI Planner alone and the AI Planner with GP. Note: the Planner1, GP1+Planner2, and GP2 rows cannot be compared because they have different GP and AI Planner parameters. This benchmark is not a comparison between the two search algorithms. Highest speedup is in **bold**

Setting	Base	Pathfinding caching	Multithreading	Both
Planner1	401.46 ± 3.23	120.96 ± 2.53 (x3)	18.23 ± 1.23 (x22)	15.68 ± 0.68 **(x25)**
GP1 + Planner2	30.85 ± 0.12	27.71 ± 0.38 (x1.1)	11.59 ± 0.06 (x2.7)	10.36 ± 0.17 **(x3)**

3.3.2 Execution Speed: Three Prongs Scenario

We run benchmarks on two scenarios: one to benchmark the AI Planner alone and AI Planner with GP, and the other to benchmark GP alone. For both scenarios, let $max(|C|) = 3$, and $|P| = 4$. We measure mean speed in seconds with and without pathfinding caching and multithreading.

Benchmark for AI Planner and AI Planner with GP Each team has four agents of one type and ten agents of another. The "Blue" team is split into two groups: the "main force" attacks "Red" team agents when they approach the city, and the "reinforcements" attack the first "Red" team agent spotted. Once that agent is defeated, the "reinforcements" will continue to attack other "Red" team agents until the "Red" team is defeated or all the "reinforcements" are defeated. Waypoints and agents are located in different spots than as detailed in Sects. 3.1 and 3.2.

Results for AI Planner and AI Planner with GP With both optimizations, the AI Planner alone receives a 25x speedup, and the AI Planner with GP receives a 3x speedup; see Table 3. Note that the Planner1, GP1+Planner2, and GP2 rows cannot be compared because they have different GP and AI Planner parameters. This benchmark is not a comparison between the two search algorithms.

Benchmark for GP All parameters, except C and P, are the same as described in Sect. 3.1.

Table 4 Average speed (seconds) and speedup using various optimizations (pathfinding caching and multithreading). We look at these measurements for the GP alone. Note: this table should not be compared to Table 3 because this benchmark is run using a different scenario. Highest speedup is in **bold**

Setting	Base	Pathfinding caching	Multithreading	Both
GP	209.99 ± 31.67	32.35 ± 2.72 **(x6.5)**	51.02 ± 1.73 (x4.1)	39.96 ± 1.16 (x5.3)

Results for GP We find that GP receives a 6.5x speedup with pathfinding caching alone but only a 5.3x speedup with multithreading and pathfinding caching (Table 4).

4 Discussion

The Four Prongs experiment shows that GP alone performs the best of the other search algorithms on average. The AI Planner with GP can perform better than the AI Planner alone, but not consistently. The AI Planner with GP sometimes scores worse than random search. One possibility is that, in the Four Prongs scenario, the benefit gained from the AI Planner is not always enough to offset the decrease in the total number of individuals generated. One thing to consider is that the performance of the AI Planner depends on the order in which plans are generated. The AI Planner in ABL-Unity3D performs a typical depth-first tree-based search algorithm.

We also show that the optimizations implemented in ABL-Unity3D can improve speedup to 25x for the AI Planner alone, 3x for the AI Planner with GP, and 6.5x for the GP alone. Though we find that multithreading improved the speed of the AI Planner alone and the AI Planner with GP relative pathfinding caching only, it did not do so for GP. In addition, multithreading alone did provide a 4.1x speedup compared to the base GP. One possibility is that our implementation can be improved. Another possibility is that the particular scenario we looked for the GP alone does not benefit from multithreading in addition to pathfinding caching; thus, the overhead of the multithreading outweighed any benefit gained. For this reason, ABL-Unity3D allows the user to disable multithreading. It is also important to note that pathfinding caching is viable only when there are no moving obstacles in the terrain. Otherwise, the caching will be invalid. If so, the user can still use multithreading to provide a speed increase. Lastly, speedups gained through these optimizations could be dependent on the scenarios tested.

5 Future Work

We presented a GUI-driven and efficient Genetic Programming (GP) and AI Planning framework designed for agent-based learning research and some benchmark results. In the future, we will examine why the AI Planner with GP performs inconsistently and does not perform as well as GP alone. One approach could be to implement the AI Planner search algorithm differently. We will also investigate why multithreading with pathfinding caching performs slower than only pathfinding caching for GP. In addition, we plan to investigate more complex GP primitives. We also plan to conduct human interaction tests to determine how user-friendly ABL-Unity3D is. Furthermore, we plan to implement features that provide more capabilities to the user. For example, the AI planner generates plans before running the simulation, but the AI Planner cannot react on-the-fly to unexpected enemy behavior. To improve this, we will implement replanning [15]. We also want to add a GUI for generating experiment statistics and visualizations for GP and AI Planning experiments. In addition, we hope adding a GUI for defining new primitives will reduce coding for domain-specific GP experiments and thus improve user-friendliness. Finally, we plan to implement co-evolution [16] of "Red" and "Blue" agents.

Acknowledgments This research was, in part, funded by the U.S. Government. The views and conclusions contained in this document are those of the authors and should not be interpreted as representing the official policies, either expressed or implied, of the U.S. Government.

References

1. Guerci, E., Hanaki, N.: Learning Agent and Agent-Based Modeling, pp. 1765–1766. Springer US, Boston, MA (2012). https://doi.org/10.1007/978-1-4419-1428-6_545
2. Abar, S., Theodoropoulos, G.K., Lemarinier, P., O'Hare, G.M.: Agent based modelling and simulation tools: a review of the state-of-art software. Comput Sci Rev **24**, 13–33 (2017)
3. Gunaratne, C., Reyes, R., Hemberg, E., O'Reilly, U.M.: Evaluating efficacy of indoor non-pharmaceutical interventions against covid-19 outbreaks with a coupled spatial-sir agent-based simulation framework. Sci. Rep. **12**(1), 1–11 (2022)
4. O'Reilly, U.M., Toutouh, J., Pertierra, M., Sanchez, D.P., Garcia, D., Luogo, A.E., Kelly, J., Hemberg, E.: Adversarial genetic programming for cyber security: a rising application domain where gp matters. Genet. Program. Evol. Mach. **21**(1), 219–250 (2020)
5. Unity Technologies: Unity3d (2020). https://unity3d.com
6. Juliani, A., Berges, V.P., Teng, E., Cohen, A., Harper, J., Elion, C., Goy, C., Gao, Y., Henry, H., Mattar, M., Lange, D.: Unity: a general platform for intelligent agents (2018). https://arxiv.org/abs/1809.02627. (10.48550/)
7. Koza, J.R.: Genetic programming as a means for programming computers by natural selection. Stat. Comput. **4**(2), 87–112 (1994). https://doi.org/10.1007/BF00175355
8. Fikes, R.E., Nilsson, N.J.: Strips: A new approach to the application of theorem proving to problem solving. Artif. Intell. **2**(3), 189–208 (1971). https://doi.org/10.1016/0004-3702(71)90010-5
9. Gold, R., Haydn Grant, A., Hemberg, E., Gunaratne, C., O'Reilly, U.M.: ABL-Unity3D (2022). https://github.com/ALFA-group/ABL-Unity3D
10. Unity Gaming Report (2022). https://create.unity.com/gaming-report-2022

11. Unity Documentation. https://docs.unity3d.com/ScriptReference/
12. Quick guide to the Unity Asset Store. https://unity3d.com/quick-guide-to-unity-asset-store
13. Aron Granberg: A* pathfinding project. https://www.arongranberg.com/astar/
14. Montana, D.J.: Strongly typed genetic programming. Evol. Comput. **3**(2), 199–230 (1995). http://vishnu.bbn.com/papers/stgp.pdf
15. Georgievski, I., Aiello, M.: An overview of hierarchical task network planning (2014). https://arxiv.org/abs/1403.7426. (10.48550/)
16. Potter, M.A., Jong, K.A.D.: A cooperative coevolutionary approach to function optimization. In: PPSN (1994)

Genetic Programming for Interpretable and Explainable Machine Learning

Ting Hu

Abstract Increasing demand for human understanding of machine decision-making is deemed crucial for machine learning (ML) methodology development and further applications. It has inspired the emerging research field of interpretable and explainable ML/AI. Techniques have been developed to either provide additional explanations to a trained ML model or learn innately compact and understandable models. Genetic programming (GP), as a powerful learning instrument, holds great potential in interpretable and explainable learning. In this chapter, we first discuss concepts and popular methods in interpretable and explainable ML, and review research using GP for interpretability and explainability. We then introduce our previously proposed GP-based framework for interpretable and explainable learning applied to bioinformatics.

1 Introduction

Machine learning has already been penetrating our everyday lives. With the increasing availability of data and computing resources, machine learning has been leveraged to successfully solve problems that were once thought impossible to tackle. Machine learning is particularly powerful at learning patterns and relationships in data too complex for humans to comprehend. Such learned knowledge has been used by machines, in turn, to make decisions and predictions throughout society.

Meanwhile, increasing demand for human understanding of machine decision-making is deemed crucial for next-generation ML/AI innovation and further applications. This is particularly important for high-stakes tasks such as applications in

T. Hu (✉)

School of Computing, Queen's University, Kingston, ON, Canada
e-mail: ting.hu@queensu.ca

Department of Computer Science, Memorial University, St. John's, NL, Canada

© The Author(s), under exclusive license to Springer Nature Singapore Pte Ltd. 2023
L. Trujillo et al. (eds.), *Genetic Programming Theory and Practice XIX*,
Genetic and Evolutionary Computation, https://doi.org/10.1007/978-981-19-8460-0_4

medicine, job hiring, and criminal justice, where the decisions can deeply impact human lives. In these applications, the consequences of failed or biased decisions can be catastrophic. Therefore, machine learning models are required to provide not only accurate predictions, but also explanations on how the predictions are made.

This has inspired the emerging research field of *explainable* AI (XAI) [25]. The research aims to explain the decisions, predictions, or recommendations made by an AI system and the process through which they are made. Explainability is the foundation to build trust in machine learning models and to further deploy AI systems for automated decision-making. Concepts related to explainability include interpretability, transparency, fairness, and trustworthy.

The terms interpretability and explainability are often used by researchers interchangeably; however, they can be different based on the way we understand a model [21, 28]. A model is interpretable if we can understand *how* a decision is made by looking at the model and its parameters. We should also be able to repeat a decision-making process of this model following the mechanism we understand. A model is explainable if we can understand *why* a decision is made. A model can be both interpretable and explainable. The meaning and tasks of interpretability and explainability depend heavily on context, data, objectives, and the nature of the end-users who are trying to understand the machine decision, represented as a mathematical model that generates predictions by learning patterns in the data [1]. Interpretability and explainability increase the transparency of an AI system. If we can understand a model and its decision-making mechanism, we will be able to better assess its fairness and trustworthiness.

Interpretability requires a model to have a symbolic representation intrinsically understandable to a human. Shallow rule-based models such as decision lists and decision trees are often considered interpretable. On the other hand, explanation techniques have been developed to provide justifications for either the global pattern/relationship learned by a model or a local individual prediction. This is often implemented by assessing feature importance or comparing the prediction to be explained with existing samples from the training set.

Genetic programming (GP) is a powerful instrument for deriving a symbolic representation to capture complex relationships in data, as well as for selecting features automatically and assessing feature importance. We see great potential for GP in interpretable and explainable machine learning. In this book chapter, we first discuss concepts and popular methods in interpretable and explainable machine learning, with a focus on the use of GP for interpretability and explainability. We then introduce our previously proposed GP-based framework for interpretable and explainable learning applied to metabolomics . We discuss its methodology and web interface design using the example of studying Alzheimer's disease metabolomic data.

2 Background

2.1 Explainability and Interpretability

Many machine learning models, especially deep neural networks, have been increasingly criticized as being "black boxes" [16]. Despite their high-performing predictions, their complex architecture and enormous parameter spaces render them impossible to understand. Ensemble methods, such as random forests and gradient boosting machines, are difficult to understand too. The ensemble is made up of many decision trees. We would have to simultaneously understand how all of the individual trees work in order to understand a prediction made by a random forest. This would not be possible when the total number of trees is large.

To support the use of these deep or ensemble models, various methods have been created to explain their predictions in a way that humans can understand. These explanation methods are often secondary to the original model and are learned post hoc (i.e., after the initial black-box model is trained) [24].

Explanation methods can be categorized into *model specific* or *model agnostic*. The difference lies if an explanation technique works only with a specific type of model or if it works with a wide range of models. For instance, various feature importance measures [3] are specific to random forests and can be used to provide explanations for a random forest prediction, i.e., which features are more influential in the prediction? Techniques like Shapley additive explanations (SHAP) [22] or local interpretable model-agnostic explanations (LIME) [27], on the other hand, assess feature importance or learn a local linear model for a given testing sample based on any of the commonly used machine learning models. They are thus model agnostic.

Another way to categorize explanation techniques is whether an explanation is *local* or *global*. Partial dependence plots [8] and random forests feature importance methods provide global explanations, i.e., which features are more influential learned by a model based on the entire training data. Some methods explain why a specific prediction is made. For instance, LIME fits a local linear model given a specific example data point and explains its prediction [27].

In addition to the techniques introduced above, there has been an explosion of explanation methods specifically developed for deep learning [29], as a result of their unprecedented popularity and the push for their wider applications. These are often post hoc model-specific explanation methods that leverage the structure of the trained neural network.

Criticism has been raised on explainable machine learning. Rudin [28], for instance, argues that for high-stakes decision-making models, we should stop creating explanations for black-box models, but instead use interpretable models. These provided additional explanations replicate some behavior of a black box that may or may not be accurate. Moreover, they refer to an understanding of how a model works, as opposed to how the world works. They provide a deceiving trust for black-box models. It is a myth that there is necessarily a trade-off between accuracy and interpretability. She suggests that we should question the use of black-box models in the first place and develop domain-specific interpretable models instead.

Descriptive definitions of interpretability have been proposed but there is no widely accepted formal or mathematical definition as of yet. It has been defined as the degree to which a machine's output can be consistently predicted [16]. From a more psychological and social point of view, it has been defined as how well a human could understand the decisions in the given context [23]. Both definitions indicate a model is interpretable if its working mechanism can be understood and its decision-making process can be consistently reproduced by a human.

Designing interpretable models is more challenging, compared with training deep black-box models since it entails significant effort in computation and domain expertise [28]. Models need to be tailored to a specific application and domain expertise needs to be incorporated into the definition of interpretability. Rule-based models and decision trees are often considered intrinsically interpretable, but they must be carefully designed and optimized to be accurate and compact for humans to understand.

Interpretability is only the first step toward creating trustworthy and ethical machine learning for social good. An interpretable model does not warrant a trustworthy and ethical model. We argue that for high-stakes applications, models should be utilized to capture and visualize the complex relationship learned in the data, as opposed to making decisions automatically. What is learned by most machine learning models are correlations not cause-and-effect relations. Models are also simulations of real world where messiness and complexity are abstracted. How a machine prediction is translated to a real-world decision needs caution and debate.

2.2 Genetic Programming for Explainable and Interpretable Learning

Genetic programming (GP) is a powerful modeling instrument with symbolic representations for predictive models, automatic feature selection , and the potential of optimizing multiple objectives. GP seems well suited for either providing explanations through feature selection/extraction, or evolving compact and interpretable models directly. In this section, we discuss GP in the context of explainability and interpretability and review research conducted on this topic.

There are recent studies using GP to provide explanations for other machine learning models. Evans et al. [5] used GP to optimize decision trees that translate the knowledge learned by any black-box classifiers. Two objectives were defined for this GP, maximizing the ability of reconstructing the predictions of a black-box model, as well as minimizing the complexity of the decision trees. The reconstruction ability is measured as the cross-validated prediction differences. The complexity of a decision tree is measured as the number of splitting points in the tree. NSGA-II [4] was employed for deriving a Pareto front of non-dominated trees. The experiments chose black-box models including random forests, gradient boosting, and deep neural networks. Testing data included binary and multi-class classification problems with both numeric and categorical features. The results reported statistically equivalent testing accuracy but considerably simpler models evolved by GP, comparing with Bayesian rule list and decision trees constructed through traditional methods.

Instead of using GP to create a secondary, global model to replicate the predictions of a black-box model, Ferreira et al. [6] proposed a GP-based method to fit a local explanation model for a given input example. Given a sample input to be explained, the method first generated a set of neighboring data points around the input based on a multivariate Gaussian distribution and then used a GP to evolve symbolic expression trees that best capture the predictions of a pre-trained black-box model on the generated neighboring data points. The authors tested on both classification and regression datasets, and reported promising results where GP was able to evolve simple and accurate algebraic expression trees to explain a prediction.

More studies can be found where GP was used to directly evolve interpretable models rather than providing secondary explanations to black-box models. Hein et al. [9] proposed a GP for reinforcement learning. A GP was used to autonomously learn policy equations for control problems from pre-existing default state-action trajectory samples. In their study, GP evolves symbolic expression trees as reinforcement learning policies. Upon applications to the mountain car, cart-pole balancing, and industrial benchmark problems, it was reported that GP was able to evolve interpretable policies of similar control performance to non-interpretable policies learned by commonly used black-box models.

In an application for the early diagnosis of Parkinson's disease through handwriting analysis, Parziale et al. [26] addressed the issue of interpretability given the high stakes of a decision in this context. The authors compared different ML techniques on both classification accuracy and interpretability, and reported that Cartesian genetic programming (CGP) was able to produce high-performing models with interpretable, explicit rules.

In addition to directly evolving interpretable models, GP is a powerful tool for feature selection and feature extraction/construction [17]. Studying features and their roles in prediction is closely related to explainable and interpretable learning. A decision can be explained through assessing the most important features in the prediction. Feature selection allows us to pick the most relevant features for model training. Feature extraction/construction allows us to build higher level condensed features for model training. Both techniques can contribute to obtaining compact, interpretable models. GP has been utilized to evolve latent features for classification [20] as well as for dimensionality reduction and visualization [19].

Compact models are innately more interpretable. There has been decades of work dedicated to the simplification of GP models, with a focus on trees. This is the well-known problem of "bloat" control in GP. Bloat control or tree simplification techniques can be readily used for interpretability. A recent review on this topic can be found here [15].

3 An Explainable and Interpretable Learning Algorithm for Metabolomics

In this section, we summarize our research on employing linear genetic programming (LGP) for the design of an explainable and interpretable learning method for metabolomics research. The framework is termed systems metabolomics using interpretable learning and evolution (SMILE) [30], where an LGP algorithm is designed to train compact genetic programs (as classification models) and a web interface is developed to allow a user to identify the most influential predictive features, to understand the interactions among features, and to interpret the model prediction.

3.1 Method and Data

In our research, a classification model is represented using a linear genetic program [2]. Compared with the traditional tree-based representation, linear genetic programs can be more expressive and efficient at search, a result of its genotype to phenotype mapping [10, 11, 13, 14]. A linear genetic program is comprised of a sequence of instructions. Each instruction is either an assignment or a conditional statement. Registers are used to hold input variable values (input registers), as well as to store intermediate calculation results (calculation registers). A specified calculation register, i.e., r_0, serves as the output register. A linear genetic program essentially maps a set of input variables to an output, which can be further converted to a binary classification result through Sigmoid transformation [12].

We use this LGP algorithm to study a metabolomics dataset for Alzheimer's disease (AD) [31]. Metabolomics is the high-throughput study of small-molecular metabolites, the footprints of biochemical processes in metabolism. The direct link between metabolism and cell/organism phenotype in health and diseases makes metabolomics an essential methodology for understanding and diagnosing disease development and progression [18]. AD is the most prevalent age-related dementia characterized, at the late stages, by the dysfunction and loss of synapses and eventual neuronal death induced by an accumulation of senile plaques and neurofibrillary tangles in the brain. The symptoms of AD include memory loss, difficulty completing familiar tasks, and personality changes. AD is a progressive neurodegenerative disease, however, causes of AD are still not fully understood. Recently, dysfunction of many metabolic pathways has been outlined as part of AD development and progression. This suggests the potential of metabolomics for a better understanding of AD and its progression [7].

LGP is used to evolve compact classification models and to identify the most influential metabolites and their interactions in association with AD. Moreover, we develop a web application with a graphical user interface that can be used for easy analysis, interpretation, and visualization of the results.

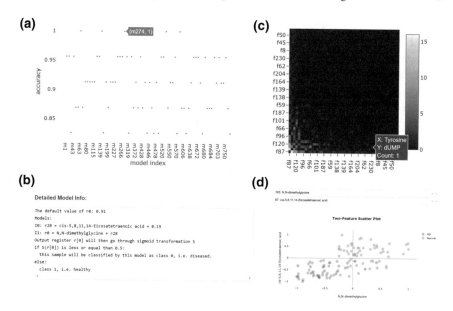

Fig. 1 Interpretable learning results of SMILE on the AD metabolomic data. All sub-figures are generated by SMILE web interface. **a** Model accuracy graph shows the testing accuracy of all predictive models that use a chosen feature. This graph, for instance, shows the testing accuracy of models that use *cis-5,8,11,14-Eicosatetraenoic acid* as an effective feature. **b** Detailed model figure provides symbolic representation of a selected model, by clicking a point in (**a**). **c** Feature pairwise co-occurrence graph represents the frequencies using a heat map, and provides the metabolite names though mouse hover. **d** Two-feature scatter plot depicts the distribution of the two chosen metabolites' concentrations. The metabolite pair shown in this figure has the highest co-occurrence frequency in the AD metabolomic data

3.2 Web Interface

We develop a web interface, https://smile-mib.cs.queensu.ca, for interpreting and visualizing the analysis results using our LGP algorithm.

First, a testing accuracy filter is provided for the user in order to include only the best performing evolved models among all collected final evolved models by running the LGP algorithm independently for 1000 times.

There are three modules for the result interpretation and visualization. The first module is *Feature Importance Analysis*. Users can decide to investigate LGP models with a specified number of effective features. Then, features are ranked based on their individual occurrence frequencies and shown in the "Feature Occurrence" graph. Clicking a feature of interest in this graph will show all LGP models containing that feature in the "Model Accuracy" graph (Fig. 1a). Further selection of a point in this graph will show its represented model in "Detailed Model Info" panel (Fig. 1b). This allows users to investigate and interpret a selected predictive model based on its testing accuracy and metabolite features involved.

Upon selecting the "Pairwise Co-occurrence Analysis" panel (Fig. 1c), users can see a heat map of "Feature Pairwise Co-occurrence", which shows all the pairwise co-occurrence frequencies in the selected LGP models. Moreover, users can manually choose a pair of features to see their distributions in diseased cases and healthy controls in "Two-Feature Scatter Plot" (Fig. 1d).

The second module is *Co-occurrence Network Analysis*. Users can visualize a network of the top most common metabolite pairs. In this graph visualization, a node is a metabolite and an edge links two metabolites if their co-occurrence frequency is above the top threshold. The node size is proportional to individual feature's occurrence frequency. The edge width is proportional to pairwise co-occurrence frequency, which is also labeled on each edge.

Users can also investigate a metabolite/feature of particular interest. The third module of SMILE is *Search a Feature*. This module allows users to enter the name of a specific feature, and will show this feature's individual occurrence frequency and its interacting features, ranked by their co-occurrence frequencies. In addition, SMILE provides a visualization of the synergy sub-network of this feature that includes all its directly interacting neighbors.

3.3 Results

Fueled by the LGP algorithm, SMILE is able to identify potential key metabolites and their interactions in association with AD. Most existing techniques provide an assessment of individual feature importance; SMILE, however, further elucidates the influence of a feature in the classification by distinguishing individual effects and synergy among pairs and multiple features. This can be particularly helpful for better understanding metabolism and AD.

In addition, linear genetic programs are innately interpretable. Through the web interface of SMILE, users can pick the most promising metabolites and investigate the best trained models utilizing them. This facilitates clear interpretation of the prediction and understanding of the data and research problems.

4 Conclusion

Explainable and interpretable machine learning is a rapidly expanding field. This is an exciting and remarkably important topic. It is extremely relevant to the next step, wider ML/AI applications, and their implications for ethics, fairness, and social considerations. Black-box models are currently widely adopted because they perform well and need little input from domain expertise. Deriving interpretable models are challenging since (1) that the features for training need to be the most relevant and sparse, (2) that the model needs to incorporate domain knowledge to be compact and accurate, and (3) that interpretability needs to be evaluated and optimized based on each application problem.

Designing and training interpretable models are problem specific and domain specific. There isn't a one-size-fits-all approach. Every application requires a clear understanding of the problem and the data, a clear understanding of what interpretability entails, and a masterful understanding of how models work and how to optimize them.

For high-stakes problems, machine learning should serve the purpose of knowledge discovery and visualization, rather than automatic decision-making. Models should be designed to best present the knowledge and patterns learned from the data. They should assist decision-makers not replace them.

GP holds great potential for interpretable learning given its ability to both evolve interpretable models and select/construct the most relevant features. We call for more research on exploring the utility of GP in this context and for next-generation, ethical AI methodology development.

References

1. Adel, T., Ghahramani, Z., Weller, A.: Discovering interpretable representations for both deep generative and discriminative models. In: Proceedings of the 35th International Conference on Machine Learning, vol. 80, pp. 50–59 (2018)
2. Brameier, M.F., Banzhaf, W.: Linear Genetic Programming. Springer (2007)
3. Breiman, L.: Random Forest. Mach. Learn. **45**, 5–32 (2001)
4. Deb, K., Pratap, A., Agarwal, S., Meyarivan, T.: A fast and elitist multiobjective genetic algorithm: NSGA-II. IEEE Trans. Evol. Comput. **6**(2), 182–197 (2002)
5. Evans, B.P., Xue, B., Zhang, M.: What's inside the black box? a genetic programming method for interpreting complex machine learning models. In: Proceedings of the Genetic and Evolutionary Computation Conference (GECCO), pp. 1012–1020 (2019)
6. Ferreira, L.A., Guimarães, F.G., Silva, R.: Applying genetic programming to improve interpretability in machine learning models. In: Proceedings of the IEEE Congress on Evolutionary Computation (CEC), pp. 1–8 (2020)
7. Gonzalez-Dominguez, R., Sayago, A., Fernandez-Recamales, A.: Metabolomics in Alzheimer's disease: The need of complementaryanalytical platforms for the identification of biomarkers to unravel theunderlying pathology. J. Chromatogr. B **1071**, 75–92 (2017)
8. Hastie, T., Tibshirani, R., Friedman, J.: The Elements of Statistical Learning. Springer (2001)
9. Hein, D., Udluft, S., Runkler, T.A.: Interpretable policies for reinforcement learning by genetic programming. Eng. Appl. Artif. Intell. **76**, 158–169 (2018)
10. Hu, T., Banzhaf, W.: Neutrality, robustness, and evolvability in genetic programming. In: Riolo, R., Worzel, B., Goldman, B., Tozier, B. (eds.) Genetic Programming Theory and Practice XIV, chap. 7, pp. 101–117. Springer (2018)
11. Hu, T., Banzhaf, W., Moore, J.H.: Population exploration on genotype networks in genetic programming. In: Proceedings of the 13th International Conference on Parallel Problem Solving from Nature (PPSN), Lecture Notes in Computer Science, vol. 8672, pp. 424–433 (2014)
12. Hu, T., Oksanen, K., Zhang, W., Randell, E., Furey, A., Sun, G., Zhai, G.: An evolutionary learning and network approach to identifying key metabolites for osteoarthritis. PLoS Comput. Biol. **14**(3), e1005,986 (2018)

13. Hu, T., Payne, J.L., Banzhaf, W., Moore, J.H.: Evolutionary dynamics on multiple scales: a quantitative analysis of the interplay between genotype, phenotype, and fitness in linear genetic programming. Genet. Program Evol. Mach. **13**, 305–337 (2012)
14. Hu, T., Tomassini, M., Banzhaf, W.: A network perspective on genotype-phenotype mapping in genetic programming. Genet. Program Evol. Mach. **21**, 375–397 (2020)
15. Javed, N., Gobet, F., Lane, P.: Simplification of genetic programs: a literature survey. In: Data Mining and Knowledge Discovery (2022). https://doi.org/10.1007/s10,618-022-00,830-7
16. Kim, B., Khanna, R., Koyejo, O.O.: Examples are not enough, learn to criticize! Criticism for interpretability. In: Proceedings of the 13th Conference on Neural Information Processing Systems (NeurIPS), vol. 29 (2016)
17. Krawiec, K.: Genetic programming-based construction of features for machine learning and knowledge discovery tasks. Genet. Program. Evol. Mach. **3**(329–343) (2002)
18. Lee, M., Hu, T.: Computational methods for the discovery of metabolic markers of complex traits. Metabolites **9**(4), 66 (2019)
19. Lensen, A., Xue, B., Zhang, M.: Genetic programming for evolving a front of interpretable models for data visualization. IEEE Trans. Cybern. **51**(11), 5468–5482 (2021)
20. Li, Z., He, J., Zhang, X., Fu, H., Qin, J.: Toward high accuracy and visualization: an interpretable feature extraction method based on genetic programming and non-overlap degree. In: Proceedings of the IEEE International Conference on Bioinformatics and Biomedicine (BIBM), pp. 299–304 (2020)
21. Lipton, Z.C.: The mythos of model interpretability. Commun. ACM **61**(10), 36–43 (2018)
22. Lundberg, S.M., Lee, S.I.: A unified approach to interpreting model predictions. In: Proceedings of the 31st International Conference on Neural Information Processing Systems (NeurIPS), pp. 4768–4777 (2017)
23. Miller, T.: Explanation in artificial intelligence: insights from the social sciences. Artif. Intell. **267**, 1–38 (2019)
24. Molnar, C.: Interpretable Machine Learning: A Guide for Making Black Box Models Explainable. leanpub.com (2022)
25. Murdoch, W.J., Singh, C., Kumbier, K., Abbasi-Asl, R., Yu, B.: Interpretable machine learning: definitions, methods, and applications. Proc. Natl. Acad. Sci. **116**(44), 22071–22080 (2019)
26. Parziale, A., Senatore, R., Cioppa, A., Marcelli, A.: Cartesian genetic programming for diagnosis of Parkinson disease through handwriting analysis: Performance vs. interpretability issues. Artif. Intell. Med. **111**, 101,984 (2021)
27. Ribeiro, M.T., Singh, S., Guestrin, C.: "Why should I trust you?": Explaining the predictions of any classifier. In: Proceedings of the 22nd ACM SIGKDD International Conference on Knowledge Discovery and Data Mining, pp. 1135–1144 (2016)
28. Rudin, C.: Stop explaining black box machine learning models for high stakes decisions and use interpretable models instead. Nat. Mach. Intell. **1**, 206–215 (2019)
29. Samek, W., Montavon, G., Lapuschkin, S., Anders, C.J., Muller, K.R.: Explaining deep neural networks and beyond: a review of methods and applications. Proc. IEEE **109**(3), 247–278 (2021)
30. Sha, C., Cuperlovic-Culf, M., Hu, T.: SMILE: systems metabolomics using interpretable learning and evolution. BMC Bioinform. **22**, 284 (2021)
31. Wang, G., Zhou, Y., Huang, F.J., Tang, H.D., Xu, X.H., Liu, J.J., Wang, Y., Deng, Y.L., Ren, R.J., Xu, W., Ma, J.F., Zhang, Y.N., Zhao, A.H., Chen, S.D., Jia, W.: Plasma metabolite profiles of Alzheimer's disease and mild cognitive impairment. J. Proteome Res. **133**, 2649–2658 (2014)

Biological Strategies ParetoGP Enables Analysis of Wide and Ill-Conditioned Data from Nonlinear Systems

Mark Kotanchek, Theresa Kotanchek, and Kelvin Kotanchek

Abstract Genetic, proteomic, and other biologically derived data sets are often ill-conditioned with many more variables than data records. Furthermore, the variables are often highly correlated as well as coupled. These attributes make such data sets very difficult to analyze with conventional statistical and machine learning techniques. The ParetoGP approach implemented within DataModeler exploring the trade-off between model complexity and accuracy enables attacking such data sets with dual benefits of identifying key variables, associations, and metavariables along with providing concise, explainable, and human-interpretable predictive models. Transparency of key variables, model structures, and response behaviors provide a substantial benefit relative to conventional machine learning and the associated black-box models. In this chapter, we describe the analysis methodology and highlight benefits using available biological data sets.

1 Goals and Objectives

Although ParetoGP via DataModeler [1] has been applied to a wide variety of domains over the past twenty years and there are a number of publications where it has been used for biological data analysis [2–6], the methodology and best practices for analyzing such data types has not been the explicit focus. We address that herein. We use four real-world data sets as surrogates for typical data seen in cell therapeu-

M. Kotanchek (✉) · T. Kotanchek · K. Kotanchek
Evolved Analytics LLC, Rancho Santa Fe, CA, USA
e-mail: Mark@Evolved-Analytics.com

T. Kotanchek
e-mail: Theresa@Evolved-Analytics.com

K. Kotanchek
e-mail: Kelvin@Evolved-Analytics.com

© The Author(s), under exclusive license to Springer Nature Singapore Pte Ltd. 2023   91
L. Trujillo et al. (eds.), *Genetic Programming Theory and Practice XIX*,
Genetic and Evolutionary Computation, https://doi.org/10.1007/978-981-19-8460-0_5

tics, immune systems and multiomics with the latter representing the integration of, metabolomics, and proteomics.

These data sets are typically ill-conditioned in the sense that there are more variables available than data records with a confounding problem that many of the variables are coupled by the biological pathways creating the data. The focus here is the methodology and general benefits of genetic programming (GP) analysis; for problem-specific insights, the reader should go to the indicated reference publications.

2 Illustration Data Sets

To motivate the methodology and workflow we will use a variety of data sets against which we have applied ParetoGP. In general, we have a limited number of patients— either human, animal surrogates, in vivo reactors—providing measurements of physical, biological, or chemical responses along with system measurements. Inclusion of genomic, proteomic, metabolomic, and cytokine data often leads to super-wide datasets with few records (patients) and thousands of variables. Longitudinal studies over time can further augment the data width.

2.1 MSC Metabolites

The immune system can be a blessing or a curse. Mesenchymal Stromal Cells (MSC) have the potential for immunomodulation with benefits for regenerative medicine as well as treating immune disorders [7]. For this data set [8], we have twenty records with an associated functional score and 8,282 metabolites—which, by definition, is a super-wide data array. Our goal is to identify those metabolites and cytokines which are most predictive for the functional score (in this case lower is better). Identifying the critical quality attributes (CQAs) will help to exploit MSC as a therapy.

2.2 CAR T-Cell

Cancer escapes detection and response by the body's immune system. Chimeric antigen receptor (CAR) T-cells are a cancer therapy approach which trains CAR T-cells to recognize cancer cells and attack them—essentially replacing broad-brush chemotherapy with a targeted immune system response. Although this approach has been successfully demonstrated, growing the requisite T-cells is a costly process. A goal of the Cell Manufacturing Technologies (CMaT) National Science Foundation (NSF) Engineering Research Center (Grant EEC-1638035) [9]) is to identify critical parameters to improve the efficacy and efficiency of production of the targeted T-cells, in this case, CD4 cells and CD8 cells [2]. CD4 cells lead the fight against infection; CD8 cells kill the cancer cells [10]. In this analysis, an initial data set (18 records and 264 variables) from a designed experiment was analyzed to identify optimal regions of process parameter space outside the original design. This Active Design-of-Experiments approach exploited the unique capability of model ensembles to extrapolate cleanly [6].

The expanded data set features 30 data records with cytokine and NMR measurements sampled at intervals during the batch process with the analytics goal to identify the key factors and develop critical quality attributes (CQAs) and critical process parameters (CPPs) for early nondestructive detection and prediction of batch quality as well as possible interventions and controls. There needs to be a balance between these two attributes; thus, the need for multi-objective optimization.

2.3 Bone Regeneration Biomarkers

After trauma, a risk factor for bone regeneration and successful healing is a dysfunctional immune system. The goal of this data set is to identify early biomarkers predictive of the long-term healing success [3]. Early intervention in such situations is important from the patient perspective. In this case T-cells and cytokines are measured at intervals. This data set is a surrogate for longitudinal studies.

2.4 Single Cell Multiomics

The original data set included a mixture of responses and liposomes from 23 patients along with single cell RNA data [4]. Since the response and liposome data are patient-specific (23 unique values), we converted the original data set dimension of 7,219 records × 1,333 columns into a 23 record × 8,452 column set by computing summary statistics for the single cell data under the rationale that since the single cell measurements fluctuate so much, a summary statistic is attractive to reduce the noise influence. The raw (noisy) RNA data is also modeled directly for comparison.

3 The Application Space

3.1 Biological Data

Biological systems can be viewed as very complex and highly coupled chemical systems. As such, they are intrinsically nonlinear and teasing insights from such systems is difficult and needs to respect the complexity as well as difficulty in collecting the requisite data to identify the key factors driving the system in order to develop robust predictive models.

Hundreds to thousands of variables and relatively few data records pose an intractable problem for machine learning techniques such as deep learning which implicitly assume every variable matters. Additionally, for biological systems, an assumption of linear or polynomial model structure is equally suspect. Correlated and coupled variables add another degree of difficulty which violate implicit assumptions of many analysis techniques.

3.2   Definitions of Success

Success in analyzing these data sets is often multifaceted with numerous objectives, such as:

- Identify the controlling variables/factors and variable combinations
- Understand the variable associations and interactions
- Develop transparent, trustable, and interpretable predictive models
- Understand response behaviors and trade-offs
- Predict optima and possibilities
- Explore competing goal trade-offs and impacts

Given the difficulty of such data sets, some measurements may be easier than others. Thus, while exploring the modeling possibilities, we also want to investigate the options for operational use. We also desire ease-of-use for the analyst and avoiding the need/temptation for the vigorous hyper-parameter tuning that seems to be endemic to much of the machine learning and statistical model techniques. Such actions introduce a risk of over-fitting and erroneous interpretations.

3.3   Data Challenges

In addition to the foundational problem of wide and ill-conditioned data sets, analysis is further complicated by correlated and coupled variables, un-captured factors and reproducibility issues across equipment, laboratories, production facilities, people, and time. Data organization and curation is a consistent problem, i.e., a data pile of 27 spreadsheets should not be confused with a data set. Surprisingly, simply choosing a target to model can sometimes be difficult since *in vitro* targets may not correspond to in vivo behavior. Missing data elements must often be addressed along with the risk of zeros being substituted for missing or unknown values.

Attempts at standardization can also be problematic with mandates for open-source code (e.g., Python) or standardized tools with limited capabilities, thus precluding innovative analysis solutions that yield deeper understanding and insight.

3.4 Biological Data Trends

The path followed by the chemical process industry is being pursued by the regenerative medicine [11] and cell and gene manufacturing [5] communities. With concepts such as systems thinking, multi-scale modeling, Industry 4.0, Quality-by-Design, digital twins, etc. being embraced, the amount of data collected and the need to efficiently and effectively convert it into actionable insight will continue to grow.

4 ParetoGP Foundations

4.1 Essential Assumptions

Our objective is predictive modeling, i.e., we have one or more known target responses that we are interested in modeling to determine what and how much to change to achieve a desired response. There are multiple approaches to predictive modeling. Each approaches makes various assumptions:

- Linear modeling assumes that the model structure is known (or reasonably approximated) and the variables are known.
- Neural networks and deep learning—which can also be viewed as a nonlinear generalization of principle component analysis (PCA) [12]—assume that every variable matters and that a sigmoid or rectifier is a reasonable basic model structure and that lots of data is available to tune the myriad of hidden parameters.
- Support Vector Regression assumes that a Gaussian located on selected data points is appropriate and that the driving variables are known.
- Gradient Boosted Trees, Random Forest, and XGBoost assume a tree-based model is appropriate. and that extrapolation or interpolation is not required.

Obviously, given their adoption, all of these can be very good modeling techniques if their implicit assumptions are aligned to the nature of the application space. In a similar vein, ParetoGP makes fundamental assumptions:

- Simple and accurate models are desirable.
- Algebraic models with relatively few (explicit) variables are appropriate.
- The proper trade-off between complexity and accuracy is an emergent property and should be determined by the data.

Beyond those constraints, the data is free to define the appropriate model form. When not constrained to a pre-defined model structure such as a second-order polynomial, evolution can be quite innovative.

4.2 Evolutionary Basics

For genetic programming, the model search is stochastic with better models awarded breeding and mutation rights. Over the course of multiple generations of evolution, models are developed. Generally, we start with an initial population of random models; however, the population can be seeded if desired.

4.2.1 Defining Model Quality

If you want to grow a better potato, you first have to be able to characterize a better potato. In a similar fashion, if we want a simple and accurate model, we need to define those terms. For purposes of ParetoGP [1], an accurate model is generally measured by the correlation of the model (R^2) to the targeted response. Since correlation is a shape-matching function, we need to scale and translate it to fit to the observed data—this is easily and efficiently achieved by a least squares fit, i.e., $y = a + b \times f(vars)$ where f is the evolved expression and a and b are the translation and scaling coefficients respectively. Since these coefficients are easily computed, we avoid making the search harder than it needs to be [13].

Model complexity is defined as a structural metric and simply the number of nodes traversed from the root node to all nodes summed together. The attraction of this metric is that it is very simple and fast to compute and a reasonable surrogate for complexity while providing a finer resolution (a larger fitness landscape) than simpler metrics such as leaf counts or model depth.

Additional objectives can be used during the model search to reward the use of fewer variables or basis functions as well as to promote novelty (e.g., model age).

4.2.2 Selecting Better Models

The models are characterized by their performance in a multi-dimensional quality space (typically complexity, accuracy, and age). For simplicity, we use these values to determine the models that contain the lowest complexity with the highest accuracy to populate the Pareto front. The Pareto front of the models identifies those models which are optimal in the sense that, there is no model more accurate while being simpler in model composition.

Rather than being computational extensive and working only with those models on the Pareto front, we choose a stochastic approach where we randomly sample a fraction (default of 10%) of the population and any models which lie upon that subset's Pareto front gets propagation rights. This Pareto tournament random sampling is repeated until we achieve the desired number of models to propagate the next generation. Although identifying the Pareto front is computationally intensive for large numbers of models, this divide-and-conquer approach has two benefits: (1) it is computationally efficient for the small number of models sampled in each tourney, and

(2) the sampling focuses the modeling selection on the knee of the Pareto front—the knee being the inflection point where increases in model complexity provide diminishing gains in accuracy. DataModeler uses $1 - R^2$ as a measure of accuracy rather than R^2 for visualization of the Pareto front knee [1]. Since knee models are most interesting from a practical standpoint, focusing the evolutionary effort on that region is appropriate. There is no need for explicit bloat control in ParetoGP due to the evolutionary preference for simplicity—— although there is a loose upper bound on the allowable complexity for safety reasons.

As an aside, models close to the Pareto front of the overall population get a free pass into the next generation to avoid losing quality results across generations—although they may lose their primacy to new models.

4.2.3 Search Time, Population Size and Independent Evolutions

Most data modeling techniques are greedy and try to identify THE model. As we shall see later, the goal of ParetoGP is to identify diverse effective models and extract insight as well as trustable model ensembles from that collection. Towards that end, we prefer to spend the computational energy on multiple shorter searches with smaller populations, analyze those results for insight and guidance and repeat the process to iterate towards a final set of variables and models.

The current best practice is to run 15–30 independent evolutions with a population size of 300 models for a duration appropriate for the data size and modeling difficulty—typically 3–30 min. Searches can be distributed in parallel to available CPUs on the computer so analyst time to achieve results are generally not onerous.

4.3 Modeling Nuances

Symbolic regression is an appropriate technology for data sets with one to tens of thousands of variables and hundreds of thousands of data records. Model search efficacy and efficiency can be boosted by appropriate adjustments as discussed in the following paragraphs.

4.3.1 Direct Model Versus Basis Set Search

As mentioned before, we search using correlation since that mitigates the need to discover the easily determined scale and translation coefficients. This is appropriate for multiplicative/monolithic models; however, if we have a system which is intrinsically additive, $y = a + b_1 f_1 (\text{vars}_1) + b_2 f_2 (\text{vars}_2) + \ldots$, then we are implicitly demanding that the evolutionary process discover both the basis functions, f_i, as well as the corresponding coefficient, b_i. Recognizing the analogy to the direct model search, we can opt to search for the basis sets directly and compute optimal coefficients as needed. This imposes an additional least-squares-fit overhead on every model evaluated but tends, on aggregate, to be beneficial for additive systems.

The results from the basis set search approach can be viewed as linearizing transforms—which can help to make symbolic regression results amenable to statisticians who can, as a result, bring into play their suite of linear analysis tools.

4.3.2 Dealing with Missing Data

Most data modeling techniques require complete numeric data with missing data handled by either deletion of offending records or imputation of missing values. The ParetoGP approach is to require that each model be purely numeric in a large fraction (default is 75%) of the data records for the variables used by each model [14]. This corresponds to a local data completeness rather than global data completeness requirement. Because the assessment is model variable combination specific, missing elements can be scattered across the data table.

Evolution can be innovative when presented with missing data. For example, Kazuhiro Iwadoh observed inclusion of a variable from a renal graft failure data set for the sole purpose of knocking out a problematic data record [15]. Although this case was obvious due to a linear term with a very small coefficient, the lesson is that the analyst needs to not chase accuracy without critical assessment of selected variables and model forms.

4.3.3 Handling Lots of Data Records and Lumpy Data

Lots of data doesn't necessarily mean lots of information, but it does impose a computational burden as we evaluate the model performance against the entire data set. Since our goal during model search is simply to determine which models are better for purposes of genetic propagation, we can use OrdinalGP [16] and, for each generation, use a different randomly selected data subset for model assessment. There are two benefits to this approach: (1) we have a computational efficiency gain due to the avoided computational load, and (2) we have an efficacy gain since models are rewarded for their generality since the fitness landscape is dynamically changing.

Data sets are often lumpy—especially for systems with closed-loop control (e.g., chemical or biochemical systems)—in the sense that data records may not be uniformly distributed across parameter space. Such data effectively overweights the lumpy regions and, implicitly, underweights other regions—which is contrary to our desire for a global model. Ideally, we would subsample the data to uniformly cover both the targeted response as well as the parameters used by the model. Unfortunately, we generally do not know which of the inputs will be used in a given model so the fallback position is to partition the response into bins (default is to use the Rice rule of $\lceil 2\sqrt[3]{n} \rceil$ which is used for histograms) and randomly sample from each bin with any deficiency achieved by randomly sampling the overall data set. This

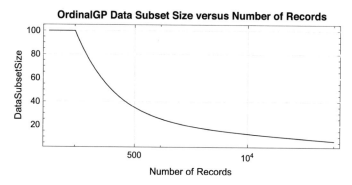

Fig. 1 Subsampling the data with each generation can provide a dramatic increase in the efficiency of model search as well as reward model generality

variant of OrdinalGP is known as BalancedGP. As implied by the default behavior shown in Fig. 1, we can be aggressive in terms of computational savings.

4.3.4 Dealing with Few Data Records

At the opposite end of the data set spectrum are wide data sets where data records are precious. This is a space where ParetoGP provides unique capabilities, as it can handle data sets with thousands of coupled variables and very few records. We do have to impose some constraints—foremost is that we want to avoid chasing phantom accuracy. Hence, the default restriction is that any model can contain no more than 20 percent of the number of records for either the number of variables or the number of basis sets. Preferably, even this upper bound will not be used in any deployed model. If we have strictly linear model, $\hat{y} = \sum_{i=1}^{N} \alpha_i x_i$, we can get a perfect model if the number of variables used is one less than the number of observations. Equivalently, N nonlinear transforms of a single variable (aka, a *basis set* of N), we also achieve a perfect model. The default behavior is intended to mitigate both of those risks.

Figure 2 illustrates the risk of spurious correlation with very small data sets. As a result, we are making an implicit assumption that the data represents reality rather than noise. Figure 2 also reinforces the need for the analyst to apply domain knowledge to validate the plausibility of the chosen model variables.

We can also use OrdinalGP and choose different data subsets for each generation. In essence, this approach is leave-one-out-cross-validation on steroids. Although it is possible to have explicit training/test/validation sets, it is generally better to exploit the diversity of the developed models and have a fully informed ensemble of models when dealing with situations where information is precious.

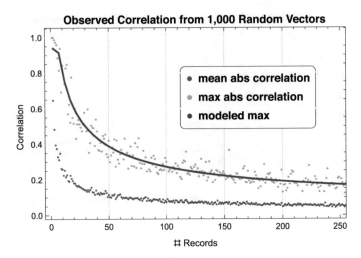

Fig. 2 The observed correlation of a random vector relative to 1,000 other random vectors is shown as a function of vector length. For length 2, the correlation is always 100% with the spurious correlation decreasing as additional data records are added. The key is that the risk of spurious correlation decreases with additional data but increases as additional variables are added—assuming the variables are random. Model transparency and human judgment are critical for a valid analysis

4.3.5 Quantized Data, Interval Arithmetic and ANOVATrim

Models developed against quantized data—and small data sets are implicitly quantized—have the problem that they are unconstrained in the interstitial region. This imposes a risk that we can have singularities in those regions which are not detected by the data assessment. An efficient, albeit conservative, means to address this situation is to use interval arithmetic [13] to kill off models which have the potential to have a singularity. This can be applied either during model development or in post-processing. This feature is disabled by default since when working with coupled variables, a singularity can be the correct behavior. However, for very short data sets, the user should probably err on the side of caution.

An ANOVA table identifies those additive top-level terms which are independent but implicitly make assumptions on the distribution (normal) of the response as well as the inputs having a balanced distribution. Although these assumptions may not be correct in practice (especially with small data sets), they are also probably conservative. Hence, out of an abundance of caution we can apply ANOVATrim to models—i.e., simply examine the ANOVA table of each model and iteratively remove top-level terms until all satisfy the specified p-value threshold (default is 0.005).

4.4 Ensembles Enable Trustable Models

During the course of an analysis, we will slowly focus from hundreds to thousands of variables to a very few which are interesting, plausible and desirable in terms of measurability and reliability. Using this focused variable set, we can have hundreds to thousands of models in the simple-but-accurate region at the knee of the Pareto front—all of which would be a viable model from a performance standpoint. Rather than choosing THE model from this candidate set, a better alternative is to choose a diverse set of models and use this ensemble as a super-model for operational use. This ensemble will have several properties:

- Models will agree near known data points; otherwise, they would not be good models.
- Models will diverge when exposed to new regions of parameter space, as they have been chosen for their diversity.

The default ensemble definition process uses the lack of correlation of error residuals as a surrogate for diversity and over-weights the knee region. Because we exploit the model diversity, as a matter of practice we want a loose definition of the knee of the Pareto front with some candidate models a little too simple and some a little too complex. The median of the ensemble becomes a good predictor—and extrapolator—and the spread of the constituent model predictions provides a measurable trust metric on that prediction.

5 The Analysis Workflow

The analysis flow when dealing with a plethora of candidate variables is fairly simple:

- Identify those variables which may be useful in predicting the target behavior.
- Iteratively model and focus on the most impactful variables considering both modeling performance and mechanistic a priori rationales as well as ease of application (e.g., ease of measurement, cost of measurement, etc.).
- Once interesting variable sets with a few critical variables have been identified, develop more models using those variable sets.
- Build ensembles (trustable models) from the diverse accurate and simple models of the sets of interesting models.

5.1 The Role of the Data Owner

The results of ParetoGP are white-box algebraic models. Hence, in the presence of coupled and correlated input variables, the data expert can and should be intimately involved in the variable selection and model selection—as we are pursuing a path of augmented intelligence rather than artificial intelligence. The situation of a few data records places an additional onus on the analyst to confirm that selected variables are plausible and not a situation of spurious correlation.

At the same time, since the data is being allowed to speak for itself in terms of variables used and model forms developed, the analyst must be willing to listen, observe, and learn.

5.2   Model Evolution

Each model search typically begins with a random and different collection of models, and follows its own stochastic path as is illustrated in Fig. 3. During model search, we want to execute a reasonable number of independent searches and let each search run for a reasonable number of generations. Our goal is to be pragmatic and identify results upon which we can build to achieve useful solutions.

As such, the evolutionary engine can be viewed as an automated hypothesis generator/refiner whose only constraint is a preference for simplicity and accuracy—presumptions of model form or a need for linearity are not part of the criteria. However, if a system can be accurately modeled as a linear system, such models will automatically emerge in the search process. Truly, the data is given an opportunity to speak for itself in terms of possible model structures.

5.3   Interesting Models—The Knee of the Pareto Front

The trade-off explored by imposing a preference for simplicity and accuracy focuses the evolutionary effort. As illustrated in Fig. 4, the models providing the lowest complexity with high accuracy are those near the inflection point. These are the set which warrant further study. In defining the interesting region, we need to resist chasing accuracy—instead, we want a mixture of slightly too simple as well as slightly too complex models. Normal practice is to keep some large fraction (50–70%) closest to the Pareto front of the models within the nominal targeted quality box to select a band along the Pareto front.

For small data sets, we probably also want to use interval arithmetic as well as apply an ANOVA p-level test to minimize the risk of unwarranted singularities or chasing outliers.

5.3.1   Model Dimensionality

One of the advantages of ParetoGP and its quest for simplicity is that we explicitly explore the number of variables needed for a model and can see the implications of chasing accuracy. Figure 5 shows that although six variables can provide the

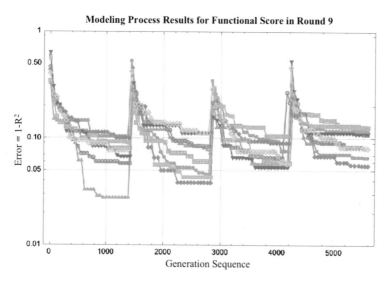

Fig. 3 Here we are looking at the trace of the highest accuracy model for 32 evolutions for the MSC metabolite data (8,282 allowed variables). Each evolution follows its own path from the initial set of randomly generated models with different variables and demonstrates continual improvement as variable combinations and models are refined. Although the initial models are not very accurate, over time better models are developed as models are refined, new variables introduced and new combinations explored. Although the default for a short (20 record) data set like this would impose an upper bound of four variables in any given model, in this case we have imposed a harsher cap of three variables per model, while all variables are eligible for model inclusion

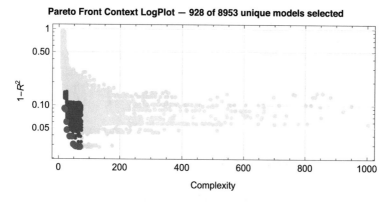

Fig. 4 Merging the results from our searches can produce thousands of models. Most are either not acceptably accurate or are inappropriately complex. However, a fraction—generally about ten percent—lie in the region at the knee of the Pareto front and represent a good balance between complexity and accuracy. Such models provide the foundation for variable as well as model selection

maximum accuracy models, we may be able to achieve good-enough accuracy with fewer variables and garner more insight in that process.

5.4 Variable Selection—Making the Haystack Smaller

Our premise in ParetoGP is that a handful of variables are sufficient for a good predictive model. There are three basic views to aid us in our focusing on the most impactful:

- *Variable Presence*—The preference for simplicity means that the interesting models feature impactful variables. Examination of these models reveal which variables are most present (Fig. 6a).
- *Variable Combinations*—The variables used by individual models or groups of models and select or combine them for additional development (Fig. 6b).
- *Variable Associations*—We can relax the specificity of the variable combinations and look at which variables associate with another and select those associations for additional investigation. Although pairwise relationships are illustrated in Fig. 6c, we can also look at triplets as well as higher-order associations.

We are helped in our search for useful variables in the biological space by the natural coupling of metabolites, cytokines, and other factors so there is a reasonable chance of uncovering relevant relationships in the stochastic search.

Our assumptions in selecting variables is very simple, but in situations where we are considering thousands of variables, we want to proceed slowly and let the computer do the heavy lifting in terms of variable selection rather than jumping directly to promising early results as seen in Fig. 7.

Although the evolutionary process is greedy in the sense that useful variables will be rewarded if they are discovered, it is not greedy in terms of only the variables with a high linear correlation to the target are allowed to participate in the model development process. The implications of this are illustrated in Fig. 8.

5.5 MetaVariables and Basis Functions

The explicit algebraic expressions developed during symbolic regression can be mined for MetaVariables, i.e., the small building blocks that are discovered and, presumably due to their usefulness, propagate through the evolved population (Fig. 9). Discovered relationships can provide mechanistic insight as well as reused as predefined building blocks for subsequent rounds of model development.

Closely related to MetaVariables are the top-level additive terms used if basis search has been enabled. These can be viewed as linearizing transforms since the resulting models can easily be dropped into a linear statistics framework. In Data-

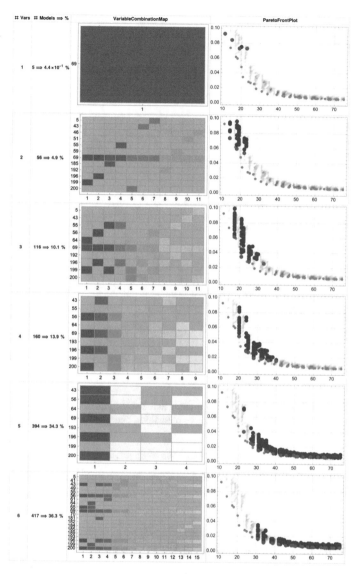

Fig. 5 The search for simplicity means that we can look at the trade-off on the number of variables (model dimensionality) versus accuracy and complexity. Here we examine the models developed using the CAR-T cell dataset where all of the process and day 6 variables were allowed in the model search (30 data records and 53 variables). Due to the default requirement that each model variable be supported by five data records, the maximum number of allowed variables in any given model was capped at 6. In this example, we obtain good models ($R^2 \geq 90\%$) with just one variable and determine that four variable models are able to provide most of the possible accuracy without adding additional complexity

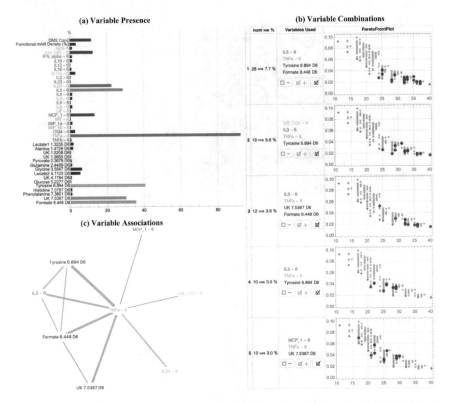

Fig. 6 Three approaches to variable for examining and focusing variable selection include identifying: **a** the variables present, **b** the variable combinations, and **c** the 2-way variable associations

Modeler, this attribute is leveraged when ANOVATrim is applied as a model filtering step.

5.6 Model Selection and Ensemble Definition

To this point in the analysis workflow, we have looked at populations of models—and, presumably, iteratively focused the allowed variable set to a small number of meaningful and useful inputs. Although the insight gained in this process can be useful, capturing the full value of predictive modeling requires a predictive model. The problem with any empirical model is one of trust that the proper variables have been chosen and the model form adequately captures the targeted system behavior. As mentioned earlier, we prefer ensembles as a way to hedge our bets. The other recommendation is to not chase accuracy by adding variables and complexity. To illustrate the latter point, consider models developed against the bone regeneration

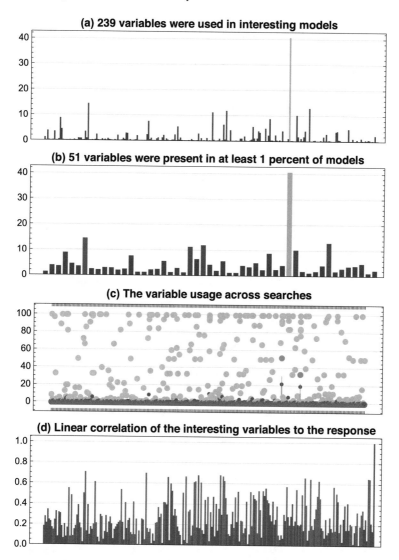

(a) 239 variables were used in interesting models

(b) 51 variables were present in at least 1 percent of models

(c) The variable usage across searches

(d) Linear correlation of the interesting variables to the response

Fig. 7 The model set used in Fig. 4 is the result of 64 independent evolutions each of approximately 200 generations. Hence approximately 4×10^6 models (not variable combinations) have been evaluated. There are 5×10^{11} unique triplets of the variables. We are truly looking at a needle-in-a-haystack problem. In **a** we see that 3% of the available variables appeared in any of the interesting models. In **b** we see only 20% of those were in 1% (~10) of the selected models. In **c** we look at the fractional presence of variables across the 52 searches which contributed to the interesting models. Despite the support of novelty and innovation, the founders effect coupled with different initial random models allow some variables to emerge and dominate the quality model results of a particular search. In **d** we show the selected variables were not chosen based upon linear correlation to the response. In fact, m5098 which was in over 11% of the models only has a 3.7% linear correlation to the target

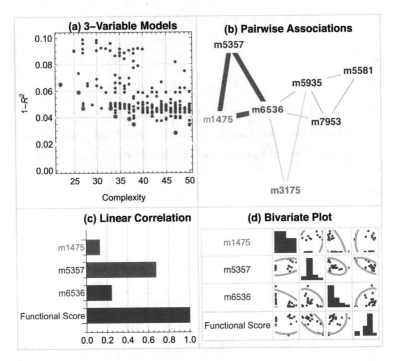

Fig. 8 From the original 8,282 variables, ten high-performing variables were identified. The top performing 3-variable models ($R^2 \geq 90$, complexity ≤ 50) were ANOVA trimmed (**a**) and variable associations (**b**) were identified. The strength of the variable associations is denoted by the line thickness. Despite the strong preference for one trio, a totally distinct set also yields quality models. We see that the variables were not chosen based upon their linear linear correlation to the response as seen in (**c**). Additionally, from the bivariate plot we can see that chosen variables cover the design space (**d**)

MetaVariables from 29 Model Searches

Rank	# models	MetaVariable	# Evolutions	% Evolutions	Max Count	Max %	Mean %
1	71	Ethanol 1.1795 D6 IL15 − 6 LIF − 6	16	55.2	16	100.0	37.3
2	48	$\dfrac{1}{OSM - 6}$	14	48.3	16	100.0	34.8
3	46	$\dfrac{IL2\ Conc}{OSM - 6}$	12	41.4	16	100.0	29.6
4	32	IL2 Conc OSM − 6	8	27.6	15	100.0	15.3
5	55	Ethanol 1.1795 D6 IL15 − 6	13	44.8	14	100.0	31.8
6	15	$\dfrac{1}{IL2R - 6}$	4	13.8	11	100.0	5.4

Fig. 9 MetaVariables are functional building blocks in developed models. Here we look for those simple structures which have been discovered and reused during the evolutionary search. These relationships can provide insight into mechanisms

data using week 9 information to predict the final bone volume shown in Fig. 10. Any model is limited to no more than four variables since we only have twenty records available; however, we can also achieve reasonable accuracy with fewer variables—albeit, at a slightly lower accuracy.

In Fig. 11, we look at the response plots of the bone regeneration ensembles at a specified reference point (in this case, one of the data records). By moving around parameter space, we can visually and numerically explore the impact of modifying each variable. If we have included a diverse set of models in our ensemble (aka, a loose definition of the knee of the Pareto front so some are under-fitted and others are over-fitted) then the constituent models will diverge when asked to extrapolate into new regions of parameter space—effectively providing a trust metric on the ensemble prediction.

ParetoGP ensembles tend to extrapolate reasonably well. To illustrate this, consider Fig. 12 where we have built models using only the process parameters in a 18 record dataset. We applied those models to the expanded data. Moving outside the nominal data range by +50% in two of the three variables is extremely aggressive, but the ensemble does a tenable good job of extrapolation while also flagging the predictions as suspect due to the constituent model divergence.

5.7 Data Cubes and Summary Statistics

The datasets considered for predictive analytics are, typically, record-oriented with each record corresponding to an observed target response. In some situations, the data can more accurately be considered to be three-dimensional with multiple values corresponding from a single measurement. To illustrate, we might have a time-series associated with a given sensor or in the case of the the single-cell multiomics, RNA measurements from individual cells of 23 patients so the nominal data set size of 7,219 records by 1,333 columns is deceptive. The RNA expression is highly variable so, as illustrated in Fig. 13a, the nominal model accuracy is not very high—it is a very noisy measurement. To take out the noise we deleted the entries for each variable which were not expressed and calculate summary statistics for the ∼300 measurements associated with each of the 23 patients.

The ability to handle wide data sets is very powerful. The example used in Fig. 13 also illustrates the risk of chasing model accuracy since a three-variable model would not have had the proper level of data support despite passing a variety of criteria. Examining the data of the selected variables is always appropriate.

5.8 Trade-Off Analysis

To this point, we have focused on a single target response. Often, we have multiple objectives which must be balanced for system optimization. Although the most

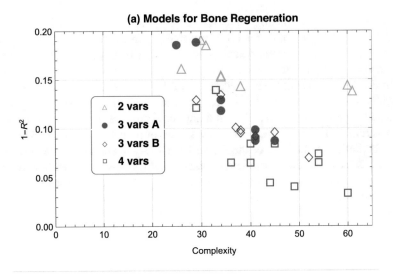

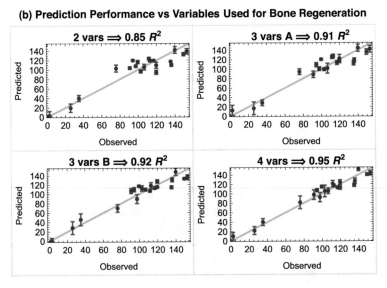

Fig. 10 Given only twenty data records available for modeling, the maximum number variables per model was capped at 4. An overlay of the Pareto front of four ensembles provides insight on the impact of the additional variables on the accuracy-complexity trade-off (**a**). Ensemble prediction plots (**b**) address whether there is a practical difference between the performance of the ensembles. Endlessly pursuing accuracy by adding more variables and complexity does not necessarily provides continuous improvement in understanding on practical performance

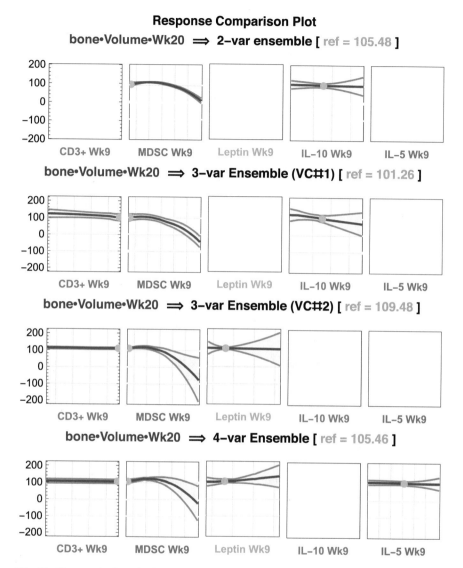

Fig. 11 Here we look at the bone regeneration behavior with each plot showing the effect of changing the associated variable while holding the others at the reference point (green dot). The key takeaway should be that the common variables across models shift similarly but the constituent models diverge (yellow envelope) when asked to operate in unexplored regions of parameter space

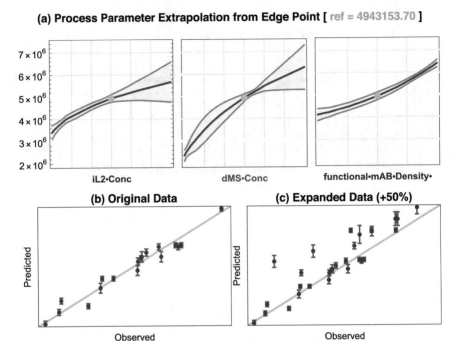

Fig. 12 Initial analysis of the original DOE data for the CAR-T dataset indicated that regions outside the nominal data range would be desirable (**a**) for further optimization. Additional data was collected in the extrapolated region focusing where the constituent models diverge. The process parameters models performed well against the original dataset (**b**) and remarkably well when asked to extrapolate +50% relative to the observed data range in two of the three variables

predictive variables may differ, we can explore the modeling options for each of the targets and exploit possible variable substitutions to identify an acceptable common set as illustrated in Fig. 14. Ideally, the chosen set of parameters will be controllable rather than observed. However, this would present new areas for data exploration by seeking the relationship between the observed parameters and the controllable parameters.

6 Competing Technologies

Unfortunately, for the very wide data sets used here, the classic machine learning techniques as well as statistical fit-regularization approaches (e.g., LASSO) struggle with the problem of too many variables with an insufficient number of data records. Iterative approaches to Partial Least Squares Regression (PLSR) is generally the most robust; however, the presumption of linearity generally precludes the variable focusing coming from ParetoGP. The required hyperparameter tuning for the other machine learning techniques should inspire caution due to the risk of overfitting generated models based on limited data records.

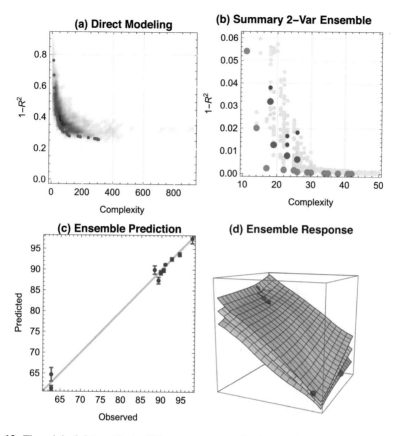

Fig. 13 The original data set had $\sim$300 measurements of each of RNA expressions for each of 23 patients. The noisy nature of the measurements made modeling difficult **a**. In **b** we use the median, interquartile range and max for each variable to produce a $23 \times 1{,}941$ input matrix and get extremely accurate models despite restricting to a limit of three variables and three basis sets and requiring each model pass both ANOVA and interval arithmetic criteria. An ensemble was formed from a 2-variable subset of the models—the individual performance is shown in **b**—and prediction performance shown in **c** with the ensemble prediction and divergence shown in **d** along with the data records. Notice that the ensemble divergence increases away from the observed data points. The key observation here is that the 23 data records only had 10 unique records so it would have been inappropriate to use more than two variables. The best practice is that, if a model is unreasonably accurate, it and the data should be scrutinized harshly

ParetoGP also naturally explores alternative variable combinations rather than being greedy in variable selection. The ability to develop diverse models and combine them into ensembles to effectively provide a predictive model with a trust metric is also critical.

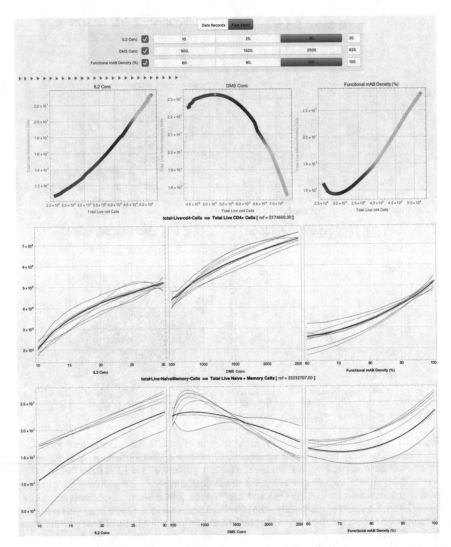

Fig. 14 The trade-off plots explore the effect of changing model variables from the reference point (green dot) from the minimum input variable value (red) to the maximum (green). Below the individual response plots are shown including the individual model trajectories of the ensemble along with the ensemble prediction in blue

7 Conclusions

Herein, we have described a workflow appropriate for biological data analysis and illustrated the associated thought process and analysis tools using a variety of real biological data sets.

Multiomics biological data is intrinsically ill-conditioned with wide data sets comprised of correlated variables. ParetoGP's exploration of explicit algebraic models rewarding simplicity as well as accuracy provides a mechanism to identify driving factors and insights into biological pathways and couplings. Combining diverse simple-but-accurate models enables robust prediction as well as a trust metric on those predictions to detect extrapolation into new regions of parameter space.

From an analyst viewpoint, the developed whitebox models with just a few variables provide clarity without requiring extensive hyperparameter tuning. Domain knowledge is important in choosing the most desirable variables for inclusion in the deployed models since multiple variable combinations are often effective predictors, and with some combinations being easier and/or cost effective to measure, monitor, and control. Finally, the risk of spurious relationships when working with very limited data sets is always a possibility; hence, the importance of review and validation by domain experts.

8 Research Opportunities

Coevolution of variable sets effective against multiple targets or a viable alternative is something that is needed since the current approach is too labor-intensive. Quantized inputs and responses is also a risk which we attempt to mitigate via the use of interval arithmetic; however, response behavior and stability in the interstitial spaces remains a concern.

Since fully exploring the search space is not practical, it may be desirable to use mutual information metrics on discovered variable sets to suggest alternatives which might also be explored for their viability.

Acknowledgements We would like to thank the leadership and staff of the National Science Foundation Engineering Research Center for Cell Manufacturing Technologies, including Krishnendu Roy, Nathan J. Dwarshuis, Maxwell B. Colonna, Valerie Y. Odeh-Couvertier, Wandaliz Torres-Garcia, and Arthur S. Edison, for their contributions in preparing and providing the CAR-T cell datasets. We thank Facundo M. Fernandez, Alexandria R. Van Group and members of the Marcus Center for Therapeutic Cell Characterization and Manufacturing (MC3M) staff, including Pallab Pradhan, Paramita Chatterjee, Carolyn Yeago, Andrew Marmon, and Annie Boules-Welch, for their contributions and support in providing the MSC datasets. We thank the University of Oregon's Guldberg Musculoskeletal Research Lab, including Robert E. Guldberg, Albert Cheng, Casey Vantucci, and Kelly Leguineche for their contributions and support in providing the bone regeneration dataset. We would also like to thank Kazuhiro Iwadoh for his insights with issues facing other machine-learning algorithms.

References

1. Evolved Analytics LLC. DataModeler. Evolved Analytics LLC, Rancho Santa Fe, USA, version 9.5 edition, 9 2021
2. Dwarshuis, N.J., Song, H.W., Patel, A., Kotanchek, T., Roy, K.: Functionalized microcarriers improve t cell manufacturing by facilitating migratory memory t cell production and increasing cd4/cd8 ratio
3. Cheng, A., Vantucci, C.E., Krishnan, L., Ruehle, M.A., Kotanchek, T., Wood, L.B., Roy, K., Guldberg, R.E.: Early systemic immune biomarkers predict bone regeneration after trauma. PNAS **118**(8) (2021)
4. Pradhan, P., Chatterjee, P., Stevens, H., Marmon, A., Medrano-Trochez, C., Jimenez, A., Kippner, L., Li, Y., Savage, E., Gaul, D., Fernandez, F., Gibson, G., Kurtzberg, J., Kotanchek, T., Yeago, C., Roy, K.: Multiomic analysis and computational modeling to identify critical quality attributes for immunomodulatory potency of mesenchymal stromal cells. In: International Society Cell and Gene Therapy - Cytotherapy, vol. 23 (2021)
5. Saplakoglu, U., Kotanchek, T., Marshall, D., Roy, K., Sobecki, S.: Expert roundtable: embracing transformation: how big data, ai and digitization are changing cell and gene therapy manufacture. Bioinsights 503–518 (2021)
6. Odeh-Couvertier, V.Y., Dwarshuis, N.J., Colonna, M.B., Levine, B.L., Edison, A.S., Kotanchek, T., Roy, K., Torres-Garcia, W.: Predicting t-cell quality during manufacturing through an artificial intelligence-based integrative multiomics analytical platform. Bioeng. Transl, Med (2021)
7. Maughon, T.S., Shen, X., Shen, X., Huang, D., Adebayo Michael, A.O., Andrew Shockey, W., Andrews, S.H., McRae III, J.M., Platt, M.O., Fernandez, F.M., Edison, A.S., Stice, S.L., Marklein, R.A.: Metabololics amd cytokine profiling of mesenchymal stromal cells idefntify markers predictive of t-cell suppression. ISCT - Cytotherapy **24** (2021)
8. Fernandez, F.M.: Facundo private communications. Metabolic Data
9. NSF Engineering Research Center for Cell Manufacturing Technologies. Cell manufacturing technology **9** (2022)
10. Haldeman-Englert, C., Raymond T. Jr., Novick, T.: University of rochester medical center health encyclopedia cd4-cd8 ratio
11. Medicine National Academies of Sciences, Engineering. Applying Systems Thinking to Regenerative Medicine: Proceedings of a Workshop. The National Academies Press, Washington, DC (2021)
12. Hinton, G.E., Salakhutdinov, R.R.: Reducing the dimensionality of reducing the dimensionality of data with neural networks. Science **313**, 504–507 (2006)
13. Keijzer, M.: Improving symbolic regression with interval arithmetic and linear scaling
14. Kotanchek, M., Smits, G., Kordon, A., Vladislavleva, K., Jordaan, E.: Variable Selection in Industrial Datasets Using Pareto Genetic Programming, volume 9 of Genetic Programming, theory and practice, vo. 6, 1st edn., pp. 79–92. Springer (2006)
15. Iwadoh, K.: Private communications. Mach Learn Comments
16. Kotanchek, M., Haut, N.: Back To The Future: Revisiting OrdinalGP and Trustable Models After a Decade, volume 18 of Genetic Programming, Theory and Practice, vol. 7, pp. 129–142. Springer (2022)

GP-Based Generative Adversarial Models

Penousal Machado, Francisco Baeta, Tiago Martins, and João Correia

Abstract We explore the use of Artificial Neural Network (ANN)-guided Genetic Programming (GP) to generate images that the guiding network classifies as belonging to a specific class. The experimental results demonstrate the ability of GP to perform such a task but also the inadequacy of most of the generated images, which can be considered false positives. Based on these findings and following an approach analogous to Generative Adversarial Networks (GANs), we propose an generative adversarial model where GP replaces the traditional GAN's generator. The experimental results illustrate the advantages of this approach, highlighting the expressive power of GP, its capacity to perform online learning, thus adapting to a dynamic fitness landscape, and its ability to create novel imagery that fits the target classes.

1 Introduction

Genetic Programming (GP) approaches for image generation purposes have a long history. The seminal works of Sims [37] and Latham [42] gave rise to a widespread interest in Evolutionary Art, which became the focus of several artists and researchers (e.g., [16, 23, 24, 34, 43]), eventually reaching mainstream notoriety. With time,

P. Machado (✉) · F. Baeta · T. Martins · J. Correia
CISUC and LASI, Department of Informatics Engineering, University of Coimbra, Coimbra, Portugal
e-mail: machado@dei.uc.pt

F. Baeta
e-mail: fjrbaeta@dei.uc.pt

T. Martins
e-mail: tiagofm@dei.uc.pt

J. Correia
e-mail: jncor@dei.uc.pt

© The Author(s), under exclusive license to Springer Nature Singapore Pte Ltd. 2023   117
L. Trujillo et al. (eds.), *Genetic Programming Theory and Practice XIX*,
Genetic and Evolutionary Computation, https://doi.org/10.1007/978-981-19-8460-0_6

expression-based GP has become the most popular Evolutionary Computation (EC) approach to image generation.

In general terms, Generative Adversarial Networks (GANs) promote the competition between two models: a generator and a discriminator. These components are trained by competing in a zero-sum game where the generator and the discriminator try to optimize opposing metrics. The simultaneous pursuit of conflicting goals is a minimax optimization scenario, similar to competitive two-player turn-based games, with the generator and discriminator taking the role of the opposing players, updating their models' weights in each turn. The use of EC to address minimax optimization problems has a long history and is often addressed in competitive evolution studies [20]. Furthermore, the idea of exploring the competition between generators and discriminators for image generation purposes has also been explored [28]. However, GANs bring two novel aspects to the table that justify, to a large extent, their massive success: adopting a representation inspired by the visual cortex (e.g., convolutional layers), which arguably facilitates image generation, and using gradient descent for optimizing the models, which speeds up optimization.

Currently, several expansions of the canonical GAN model exist (e.g., [1, 46]), the robustness of training has been improved, and GAN research is thriving. However, GANs still have several shortcomings, such as mode collapse and instability, that compromise their generative performance [14].

GANs have two key advantages over GP:

- They use the parallelization capabilities of modern Graphic Processing Units (GPUs), Tensor Processing Units (TPUs), and other throughput-oriented processors alike.
- Their ability to generate new data samples from pre-existing instances, synthesizing new instances that follow the distribution of the training set. In other words, while expression-based GP tends to generate images that have a mathematical and abstract appearance, GANs can "effortlessly" mimic figurative imagery.

Considering the emergence of GP frameworks such as TensorGP [3], which is based on TensorFlow, and thus allows the seamless integration with GPU computing, we believe that it is timely to revisit GP-based image generation. A set of findings and observations support this belief, including the following:

- It is well-established [24] that expression-based GP has the expressive power to generate any given image. As such, from a theoretical perspective, there is no reason to dismiss GP from being a generic image generation approach.
- Expression-based GP has been successfully used to generate images that Artificial Neural Networks (ANNs) classify as belonging to a given class [6, 25, 29].
- Likewise, GP can easily create fake and adversarial examples (see, e.g., [6, 32]).
- Because the expressions to evolve are problem-independent, GP does not have an a priori need for domain-specific knowledge about the structure or form of the task at hand.
- Linked with the previous point, there is reason to believe that GP is less likely to match precisely the training data distribution, which may alleviate the problems

raised by training set biases and allow the generation of samples that deviate from the dataset distribution, but that still represent valid instances.

• Expression-based images are resolution-independent, allowing the creation of large-format images [37].

The viability of employing GP for adversarial image generation depends on a critical issue, what is the *de facto* expressive power of expression-based GP? While theoretical findings demonstrate that any image can be generated, reality indicates otherwise. To date, expression-based GP has failed to produce photorealistic images, letters, numbers, and many other types of images that GANs routinely create. A competitive setting implies that the generator and the discriminator change over time. Therefore, the generator will tend to face naïve, easy to fool, classifiers at the beginning of the run, which may facilitate evolution, and adversarial hardened classifiers in later stages. As such, a second research question arises, what is the impact of competition, and the dynamic fitness landscape it induces, on the evolutionary process, namely on the ability of the generator to create images that can be considered true positives.

The work presented herein is a first step toward developing EC-based Generative Adversarial Models. We begin by assessing the expressive power of expression-based GP by testing its ability to generate images classified as digits by a Convolutional Neural Network (CNN) trained on the MNIST dataset. The results highlight the ability of GP to generate such instances and also their adversarial nature. Next, we present TensorGP Generative Adversarial Network (TGPGAN), an adversarial model that couples a GP generator (TensorGP) and a standard convolutional discriminator, and that trains them online and adversarially.

The experimental results show the advantages of the approach, namely the following: *(i)* Its ability to generate a diverse set of images that increasingly resemble the real data; *(ii)* The ability of GP to tackle the dynamic and increasingly demanding fitness landscape that results from the iterative refinement of the discriminator; *(iii)* The advantages of starting with an easy fitness landscape.

The structure of the chapter follows. In Sect. 2, we perform a short survey of the use of GP for image generation, emphasizing works where Machine Learning (ML) models, namely ANNs, provide fitness, and we highlight GP-based adversarial models that precede GANs, indicating their key strengths and shortcomings. In Sect. 3, we introduce TensorGP, the TensorFlow-based GP engine that powers the generator component of our framework. Next, in Sect. 4, we present the experiments concerning image generation using an ANN-guided GP, presenting the experimental results and their analysis. A description of our adversarial GP-based generative framework follows. The experimental results obtained with this adversarial framework and its analysis are presented in Sect. 6. Finally, we draw some conclusions and indicate future research.

2 State of the Art

As previously mentioned, the work of Sims [37] popularized the use of GP for image generation. Like in canonical GP [21], the genotypes are trees that represent symbolic expressions. The images can be seen as visualizations of the output of such expressions over a given x and y interval (see, e.g., [24] for a detailed explanation of the genotype to phenotype mapping process). This approach has become known as expression-based GP and has since then been adopted by a large number of researchers for image generation [4, 16, 23, 24, 34, 43]. While several other GP-based image generation approaches exist (see, e.g., [22, 27, 33]), it is safe to say that expression-based approaches are the most common ones. A special word goes toward approaches based on Compositional Pattern Producing Networks (CPPNs) [40], such as Picbreeder [35, 36]. While these works are based on NeuroEvolution of Augmented Topologies (NEAT) [41], and thus technically different from Sims-like expression-based GP, they share enough similarities for us to consider them a particular case of expression-based GP. Due to its widespread adoption, we will focus exclusively on expression-based approaches in the scope of this paper.

An important advantage of expression-based solutions is their ability to generate an output for any value of the function's domain, allowing for an infinite level of granularity. This capability is compelling when applied to evolutionary art because it enables the synthesis of images with an arbitrary level of detail.

The work of Baluja, Pomerleau, and Jochem [4] is the first attempt to automate fitness assignment in the context of evolutionary art. They trained ANNs to learn the users' preferences, then used the trained ANNs for fitness assignment. While the training was successful according to ML standards, the authors considered the experimental results "disappointing", since the GP engine quickly found ways to maximize fitness without actually improving image quality. In the same year, Spector and Alpern [38] reported similar findings in the musical domain, highlighting the ability of GP to exploit the deficiencies of the ANN used to assign fitness, finding "shortcuts" that allowed it to maximize fitness with melodies that were undesirable and close to inaudible. As far as we know, these are the first reported instances of what is now known as "adversarial" examples.

Years later, driven by the desire to evolve representational images, Correia et al. [6, 25] employed off-the-shelf image classifiers to evolve images that they classified as members of a given class. Although some degree of success was achieved, as illustrated by the generation of several figurative images, the runs converged to false positives in most cases. Using a similar approach, Machado et al. attempted to evolve ambiguous images [29], reporting equivalent findings. More recently, Nguyen et al. [32] arrive at similar results, demonstrating that Deep Neural Networks (DNNs) are also easily fooled by expression-based GP. While the previous works tend to generate false positives, the work of Correia et al. [9] shows that it is possible to use EC to evolve false negatives, by generating human faces that are not recognized as such by a face detector (but that are effortlessly recognized as faces by humans).

In summary, the state of the art regarding ANN-guided GP demonstrates the ability and tendency of GP to find adversarial examples. This tendency has been interpreted as a lack of generative power, but we disagree. These results only demonstrate the vulnerabilities of current ANNs and the ability of GP to exploit them.

Adversarial approaches, namely co-evolutionary ones, have been employed since the early years of EC research. In the field of image generation, the work of Greenfield [17] is among the first ones. In 2007, Machado et al. [28] introduced a system that has many resemblances with GANs. Most notably, the system is composed of a generator and a discriminator, establishing a competition between them. The discriminator is an ANN trained to distinguish between images of famous paintings and images created by the generator. The generator is an expression-based GP system that attempts to fool the discriminator. Unlike GANs, the training is not performed online. Instead, once a GP run is finished, the generated images are added to the training set, and the discriminator is trained from scratch. A new evolutionary run is then started from an initial random population. This process is repeated until one of the systems becomes unable to cope. Correia et al. [7], via the EFFECTIVE framework, improve upon this model by including novelty search mechanisms, an archive that summarizes the range of images produced by the system, and algorithms to manipulate that are to be added to the training dataset. Correia et al. [8, 26] have used similar approaches for data augmentation.

The advent of GANs has revolutionized image generation, surpassing these earlier approaches. Nevertheless, the use of EC in this context is still fertile. To stabilize GAN training and improve generative performance, Wang et al. proposed Evolutionary Generative Adversarial Networks (EGANs) [44], which replace the generator with a population of generators that adapt to the environment through selection and reproduction. Costa et al. introduced the CoEGAN model [10], extending EGANs through the integration of neuroevolution and co-evolution, seeking to optimize DNNs architectures by evolutionary means.

Ha incorporated a CPPN architecture with the adversarial pipeline of a Variational Auto-Encoder (VAE) GAN [18, 19], which allows backpropagating gradients over the generator network. The proposed model is able to generate high-resolution artifacts that retain visual similarities to the original samples. Nevertheless, the model trained on the CIFAR dataset demonstrated the limitations of this approach in generating a truthful representation of rich images.

Metz and Gulrajani [31] extended upon Ha's research by using a Deep Convolutional Generative Adversarial Network (DCGAN) architecture and training different discriminators for different datasets while training the generator to various approximate distributions at different scales. Notwithstanding the impressive generative performance, the model was "unable to reach the same visual quality of existing convolution transposed models even after millions of steps of training" [31].

More recently, the work of Ekern et al. [13] further demonstrates the effectiveness of CPPN-GAN models by proposing a framework that maximizes the mutual information between the latent code and the generated images using an auxiliary network.

One of the core challenges of coupling GANs with the expression-based paradigm, as discussed in the next section, lies in the fact that isolated solutions are generated instead of a latent space, which is characteristic of GANs and other generative approaches. The organization of possible data samples by regions provided by the latent space is desirable as it allows for the conditional exploration of solutions through algebraic vector operations. Alternatives to mapping individual solutions into an organized latent space include the addition of an archive to store solutions according to their fitness and novelty [7, 30]; manifold learning algorithms to reduce dimensionality [5]; and the use of VAEs by training an additional network that learns to encode an image into a much smaller latent vector. Additionally, recent work has shown the effectiveness of local embedding techniques to perform dimensionality reduction and assess the progress and visualization of GAN training [11]. Although the summarization, visualization, and exploration of the range of solutions generated by GP is a fascinating research topic, we will not address it in this chapter.

3 TensorGP

TensorGP [3] is the GP engine that powers the Generator component of TGPGAN and is publicly available on Github.[1] The engine is based on the standard tree-based approach for representing individuals in GP and expands on it by using tensor operations to speed up the evaluation of individuals. This is achieved using the TensorFlow Python library which takes advantage of parallel computing capable hardware. Additionally, TensorGP allows for the caching of intermediate results, which accelerates the evolutionary process by avoiding the re-execution of common sub-trees.

3.1 Genotype to Phenotype Mapping

The simplest form of representing an individual in GP is through a mathematical expression, as exemplified in Fig. 1. The aforementioned tree-based approach that TensorGP follows requires an initial mapping phase from the initial expression genotype to a tree graph. However, this first transformation is only performed at the beginning of the evolutionary process when initializing the population. Next, TensorGP uses TensorFlow to internally convert the tree graph into a Directed Acyclic Graph (DAG) before calculating any output values, enabling the caching of potential intermediate results from sub-trees that appear frequently among individuals in the population. Lastly, the final mapping phase goes through the entire set of data points by querying the internal DAG to produce the phenotype.

Because the domain of fitness data points to be evaluated is fixed for all operations, the vectorization of this data is made trivial using a tensor representation. Generally

---

[1] TensorGP repository available at https://github.com/AwardOfSky/TensorGP.

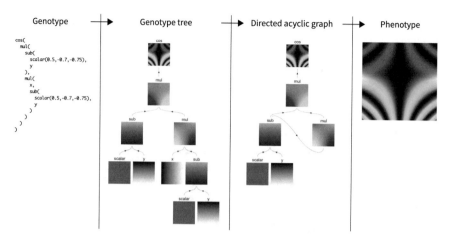

Fig. 1 Genotype to phenotype mapping phases in TensorGP

speaking, the phenotype is a tensor obtained through a composition of operators, variables, and constants that make the individual. This tensor, commonly represented as an image, will be the target of fitness assessment.

3.2 Function and Terminal Set

TensorGP provides sufficiency through the redundancy of operators by implementing a function set that goes beyond the scope of basic mathematical and logic functions. Some image-specific operators are also included to facilitate the application of TensorGP to evolve images. An example of this class of operators is the warp, which transforms an image by mapping each pixel to a different coordinate. All operators are applied to tensors and integrated into TensorGP by employing existing TensorFlow functions.

In addition to the standard protection mechanism to avoid invalid operations (e.g., division by zero), there are essential implementation details for some operators. For example, when calculating trigonometric operators, the input argument is first multiplied by π. The reasoning behind this is that most problem domains are not defined in the $[-\pi, \pi]$ range but are otherwise normalized to either $[0, 1]$ or $[-1, 1]$. This makes it so that the argument to the trigonometric operators is in the $[-\pi, \pi]$ range, which covers the whole output domain for these operators.

The default terminal set provided by TensorGP encompasses an array of coordinate tensors equal to the number of indexable dimensions as well as an operator to define scalar tensors. Coordinate tensors correspond to the variables x, y, z, ... that typically define a coordinate along one dimension, but are instead indexed for every data point in the problem domain. It is also worth noting that both the base TensorGP function and terminal set can be extended through the definition of custom operators and tensors, respectively.

3.3 Speed Gains

By seamlessly distributing computational efforts among the available hardware, TensorGP can achieve speedups that scale with the amount of parallel processing capability available. Previous experimentation with TensorGP [2] demonstrated speedups of up to two orders of magnitude compared to well-established frameworks that perform an iterative evaluation and similar vectorization-boosted frameworks, such as KarooGP [39].

The performance gains achieved by TensorGP are more pronounced in the evaluation of domains containing millions of data points. Moreover, execution on throughput-oriented architectures, such as the GPU, is also shown to increase TensorGP performance gains. These speedups make the evaluation of large sets of data points feasible, which in turn provides more information that may be useful to approximate and discover optimal solutions. Due to these characteristics, TensorGP is particularly well-suited for GP-based image generation, hence its adoption.

4 GP-Based ANN-Guided Image Generation

The feasibility of GP-based generative approaches depends on their *de facto* generative power. While it is well-established that traditional expression-based GP can theoretically be used to generate any image, it is also well-known that, in practice, finding the symbolic expression to a given target image is a complex problem. More precisely, finding a compact symbolic expression that matches or approximates a given target image is, in general, a hard problem. This raises doubts regarding the "practical" generative power of expression-based GP.

We claim that matching a target image is not a good test for our purposes. Instead, what is relevant is the ability of GP to generate images that are classified as being members of a specific class.

As we demonstrate in Sect. 2, it is currently well-established that GP and similar methods can easily fool ANNs, including the state-of-the-art DNNs. However, as far as we know, there are no studies analyzing the ability of expression-based methods to consistently and systematically generate examples classified as belonging to a given class.

This section presents a set of experiments where we use a pre-trained ANN to assign fitness guiding the GP engine. The classifier used for this task is a CNN pre-trained on the MNIST dataset [12], that achieves a test accuracy of 99.25%, and, therefore, the GP engine will aim at evolving images that are classified as digits.

Table 1 Parameterization of TensorGP for the MNIST experiment

Parameter	Value
Generations	101
Population size	50
Elitism (elite size)	1
Tournament size	5
Mutation probability	0.9
Mutation operators	Delete, insert, point, sub-tree
Crossover probability	0.7
Crossover operators	Random sub-tree swap
Minimum initial depth	2
Maximum initial depth	12
Minimum allowed depth	2
Maximum allowed depth	12
Generation method	Ramped Half and Half
Fitness metric	CNN classifier
Domain range	$[-1, 1]$
Function set	add, sub, mul, div, abs, min, max, neg, warp sqrt, mdist, sin, cos, if

It is essential to notice that we are testing the ability of the ANN and GP to work in conjunction. The ANN determines a significant part of the fitness landscape; if it fails to provide gradients to guide the search, the evolutionary engine can only do a random search. We also wish to assess the nature of the evolved images, namely if they are "exploits" or if they are believable representations of the target digits.

Although the MNIST dataset can be considered too simple for most modern ML applications [45], the fact that we are exploring largely uncharted territory recommends starting from the basics in order to establish a solid foundation. Additionally, MNIST samples are small (28×28 pixels), significantly reducing computational costs.

Table 1 summarizes the parameters used to configure TensorGP during the course of these experiments.

The function set plays an important role in the evolutionary process. For the experimentation included in this work, we consider simple mathematical and trigonometric operators along with image-specific operators such as the warp [3] operator. Additionally, a conditional operator is also included to allow for the definition of different expressions for different regions of an image and ensure that the model can generate any arbitrary image.

The fitness is the activation level of the output neuron encoding the desired class.

4.1 Experimental Results

For each digit, we perform 30 independent runs of TensorGP. Table 2 outlines the results of these runs.

As it can be observed, it is trivial to evolve images classified as the digits 2, 3, 4, 5, 7, and 8, since, for these digits, TensorGP was always able to find an image that maximizes the ANN output. Conversely, evolving 0 and 6 is significantly more challenging but still feasible. These results are not surprising; after all, evolving 28 × 28 grayscale digits should not be a complex task. At the same time, nobody ever evolved images of digits or letters using expression-based GP to the best of our knowledge.

A more detailed analysis reveals the reason for this uncanny success, in many cases (i.e., for digits 1, 2, 3, 7, and 8), the optimum solution is almost always found in the first generation of the run. No evolution is required. For digits 0 and 6, the algorithm must perform, on average, more than 30 generations to optimize fitness, but it is not always able to find the optimum. Digits such as 4 and 5 typically require a few generations to reach the optimum, but some runs demand significantly more generations than others. Finally, the evolution of images classified as a 9 follows a pattern similar to those regarding 4, although, in this case, a single unsuccessful run hinders the performance.

Figure 2 illustrates the different behaviors found during these experiments, by presenting the evolution of the fitness of the best individual during the course of the 100 generations for digits 0, 4, and 7.

Table 2 Analysis of the best results for the evolution of each digit. Columns from left to right for each digit: average fitness of the best individual out of 30 runs, standard deviation of the fitness of the best, number of runs in which the maximum value (1.0) was attained (out of 30), average generation number where optimum was found (taking only into consideration the runs where it was found)

Digit	Avg (Best)	Std (Best)	Found optimum	Generation optimum
0	0.74	0.35	17	38.71
1	0.99	0.01	30	0.00
2	1.00	0.00	30	0.13
3	1.00	0.00	30	0.00
4	1.00	0.00	30	12.50
5	1.00	0.00	30	21.17
6	0.78	0.38	22	30.86
7	1.00	0.00	30	0.50
8	1.00	0.00	30	0.00
9	0.97	0.17	29	11.34

Fig. 2 Evolution of the
fitness of the fittest
individual across generations
for different target digits.
The blue line indicates the
average of 30 runs, while the
box-whisker represents the
minimum, maximum, Q1,
Q3, median, and outliers

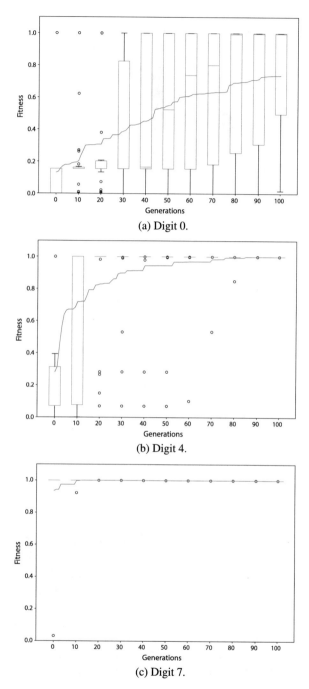

(a) Digit 0.

(b) Digit 4.

(c) Digit 7.

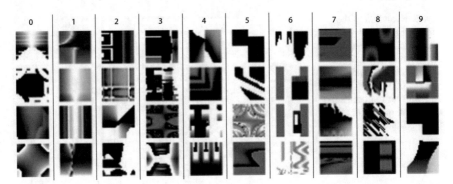

Fig. 3 Random sample of the fittest individuals evolved using ANN-guided GP. Each column corresponds to a given digit, from 0 (leftmost column) up to 9 (rightmost column). All images maximize the activation of the corresponding output neuron

Figure 3 reveals the nature of the images being evolved. As expected, the vast majority of the examples are false positives, and, unsurprisingly, even when the evolutionary search is required, GP is essentially exploiting the shortcomings of the ANN. These results are in line with those reported in previous experiments. However, they illustrate that even extremely good classifiers, remember that this particular classifier has a 99.3% success rate on the MNIST dataset, can be easily fooled. This is especially true when they are confronted with instances that are radically different from those used to train them, and that is one of the advantages of generating synthetic data without relying on a priori knowledge.

As we will see in upcoming sections, the use of an adversarial model is able to improve the classifier, making it less susceptible to adversarial attacks, and allowing the evolution of true positive instances.

5 Framework—TGPGAN

In this section, we describe TGPGAN a variation of a conventional DCGAN. Like in a traditional GAN, we have a Discriminator, which attempts to determine if a sample is genuine or fake, and a Generator, which tries to create samples that are classified as genuine. In TGPGAN, like in traditional GANs, the Discriminator is a CNN. However, the Generator is a GP system (Fig. 4).

The feedback of the Discriminator guides the Generator's evolutionary process by supplying the fitness values (similarly to the process described earlier in Sect. 4). The source code of TGPGAN is publicly available on GitHub.[2] Algorithm 1 presents an overview of the entire TGPGAN adversarial system.

---

[2] TGPGAN repository available at https://github.com/AwardOfSky/TensorGP_DCGAN.

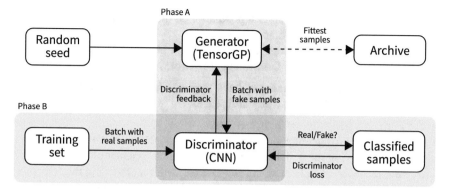

Fig. 4 Overview of TGPGAN

Algorithm 1: Training loop implemented by TGPGAN.

---

foreach *training_step* **do**

$n = round(training\_step/10) + 5$;

Generate batch of fake samples **from** GP run with n generations, and k fittest samples from last training step and y samples from archive if applicable;

Update archive **with** fittest samples;

Get batch of real samples **from** training set;

Train discriminator **with** batch of real samples;

Train discriminator **with** batch of fake samples;

Calculate losses;

The process starts with phase A. At this stage, since we are at the beginning, the Discriminator still has not been trained, and the archive is empty. As such, we create an initial random population and start a GP run, consisting of n generations, using TensorGP. The GP individuals are assessed during the evolutionary process by a forward pass of the Discriminator network. Fitness favors images classified as genuine.[3] Once the n generations are performed, the last population of the GP is passed to the Archive manager and to the Discriminator. This concludes phase A, and we enter phase B. The Archive manager compares the fitness of the individuals on the archive with the fitness of those being added, keeping the *size* fittest ones. When draws exist, the newest individuals are kept. In the current version of TGPGAN, *size* is equal to 100.

Phase B consists of updating the Discriminator, which is trained on two batches of images:

---

[3] In our case, the Discriminator has a single output neuron, and it is trained to return 1 for genuine images and 0 for fake ones. In these circumstances, fitness is equal to the activation of the output neuron.

- A batch of fake instances—the last population of the Generator;
- A batch of real instances—randomly selected samples from the training dataset. We employ standard backpropagation training coupled with the Adam Optimizer to update the weights of the Discriminator network.

This concludes phase B, taking us back to phase A. The process is, in essence, the same as previously with two key differences:

- The number of generations, n, is increased;
- Instead of using a random population, the initial population of the GP algorithm is created by copying the f fittest individuals of the last population of the previous GP run, the a fittest individuals of the archive, and then creating the remaining individuals randomly. In the experiments presented in this paper, we only copy one individual from the last population and one from the archive.

Notice that at the beginning of the process, the Discriminator has random weights. Therefore, it is straightforward to fool it. However, as the number of cycles increases, the Discriminator is exposed to an increasing number of genuine and fake samples, improving over time. Thus, unlike most GP scenarios, the fitness function is dynamic, and the fitness landscape is increasingly more complex. To deal with this increase in difficulty, the number of generations of each TensorGP run grows as the number of cycles advances. In the experiments reported in this chapter, we start with 5 generations and increase the number of generations by one every ten training steps (see Algorithm 1).

6 GP-Based Adversarial Image Generation

In this section, we analyze the experiments conducted with TGPGAN on the MNIST dataset. These results are compared with those obtained with the GP-based ANN-guided image generation approach from Sect. 4 and with a conventional DCGAN model, adapted from TensorFlow's website.[4] Moreover, we use an external CNN model to evaluate generated images and assess their recognizability.

The network used for the Discriminator of both models is a CNN-based image classifier consisting of two convolutional layers with a dropout of 0.3 and a final dense layer for flattening. Each convolutional layer has a kernel size of 5×5, a stride vector of [2, 2], and uses the Leaky Rectified Linear Unit (LeakyReLU) as an activation function.

The Generator network for the DCGAN model starts with a dense layer followed by three transposed convolutional layers used for upsampling and a Batch Normalization. The parameterization is the same as for the Discriminator network, except in what concerns the first convolutional layer, which has a stride vector of [1, 1], and the last layer, which uses the hyperbolic tangent as an activation function instead of the LeakyReLU.

The Generator of TGPGAN uses TensorGP. The relevant parameters are presented in Table 3.

---

[4] Discriminator adapted from https://www.tensorflow.org/tutorials/generative/dcgan.

Table 3 Parameterization of the TGPGAN model and GP run

Parameter	Value
Number of generation	$5 + (training_{steps}/10)$
Population size	32
Elitism (Elite size)	1
Tournament size	2
Mutation probability	0.3
Mutation operators	Delete, insert, point, sub-tree
Crossover probability	0.8
Crossover operators	Random sub-tree swap
Minimum initial depth	3
Maximum initial depth	6
Minimum allowed depth	3
Maximum allowed depth	14
Generation method	Ramped half-and-half
Archive selection	1
Archive max size	100
Fitness function	Discriminator's loss after forward pass
Domain range	$[-1, 1]$
Function Set	add, sub, mul, div, abs min, max, neg, warp, sign sqrt, pow, mdist, sin, cos, if

We trained the 10 digits of the MNIST dataset using the DCGAN and TGPGAN algorithms separately while using the defined experimental setup for 5 epochs. In this context, we consider an epoch to be a full cycle over the real samples of the training dataset, i.e. the real digits. Thus, given that the MINIST dataset has 60,000 training samples and that the batch size is 32, on average, an epoch takes 187.5 training steps (see Algorithm 1). At the end of this process, the experiments yield 20 models, each trained with its corresponding digit and approach.

6.1 Experimental Results

Figure 5 depicts the evolution of the average fitness of the best individuals of TGP-GAN during the first 500 generations of the first epoch of the overall training process along with the corresponding fitness deltas. As we mentioned previously, the main objective of the GP is to create individuals that maximize the loss of the Discriminator, i.e., evolving images that it classifies as real.

Fig. 5 Best fitness values of
the evolution of the digit 0
using TGPGAN at the start
of the evolutionary process.
Each vertical dashed line
represents the end of a GP
run and the generation of a
new fake batch. Higher is
better

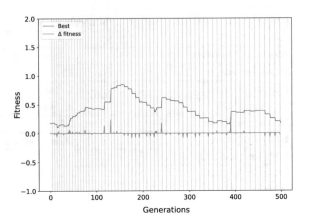

The dashed lines represent the end of a GP run and, as mentioned in Algorithm 1, whenever a run ends, the Discriminator is trained with a batch of real and the fake samples.

Thus, an increase in the best individual's fitness during the generations between the dashed lines is expected. However, since each vertical dashed line corresponds to an update of the Discriminator, it is natural that a drop in fitness accompanies a dashed line. This expected behavior is present in Fig. 5. As it can be observed, the best fitness values tend to increase between dashed line intervals and tend to drop with the lines. The increase in the number of generations can be noticed by the increase in the space between dashed lines. Moreover, the variation of the fitness of the best individual from one generation to the next in fitness tends to be small, and the most noticeable variations are associated with changes of the discriminator, and thus, of the fitness function.

Figure 6 shows the batches generated by TGPGAN during the first 450 generations depicted in Fig. 5. As it can be observed, in the first training steps, the Discriminator is easily fooled by images that share (at least to the human eye) little to no resemblance with a handwritten zero. In later stages, the images are arguably closer to a zero, with many depicting closed forms that, roughly, approximate a circle.

Figure 7 is similar to Fig. 5. However, it shows the average fitness of the best individual across the entirety of the 5 training epochs, depicting the results at a macroscopic scale. Here the dashed line represents the change of epoch.

We can observe the large fluctuations in fitness throughout the training process, which indicates that Generator and Discriminator are actively competing and that neither can surpass the other during long periods. This result contrasts with the behavior depicted in ANN-guided evolution, where the GP was quickly able to outperform the Discriminator.

As is usual in image generation, the paramount results tend to be the images themselves. Figure 9 presents random samples of the fittest individuals stored in the TGPGAN's archive.

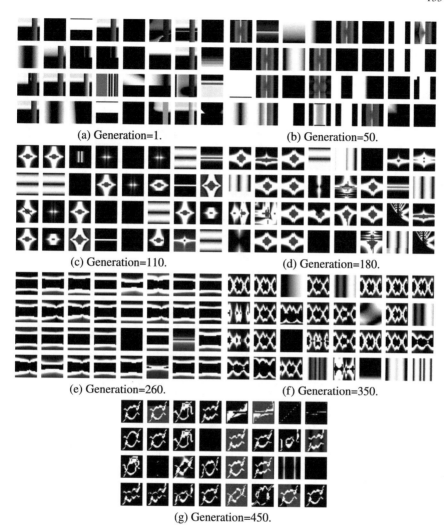

(a) Generation=1. (b) Generation=50.

(c) Generation=110. (d) Generation=180.

(e) Generation=260. (f) Generation=350.

(g) Generation=450.

Fig. 6 Sample of the evolution of the batches of images generated by TGPGAN while evolving the digit 0 during the first 500 generations (see Fig. 5)

The differences between this image, where all the digits are easily recognizable, and the false positives presented in Fig. 3 are striking. We believe this comparison shows the undeniable advantages of TGPGAN over ANN-guided GP. Indirectly, it is also a demonstration of the generative power of expression-based GP. As we can observe, it is perfectly possible to evolve convincing digits, provided that the algorithm is properly guided to the promising regions of the search space.

In addition to the comparison between TGPGAN and ANN-guided GP, we also compare TGPGAN with DCGAN. Figure 8 presents the loss of the Discriminators

Fig. 7 Best fitness values across generations for the evolution of 0's, 4's, and 7's (respectively, from top to bottom). The vertical dotted lines mark the start/end of GP run in TGPGAN. Higher is better

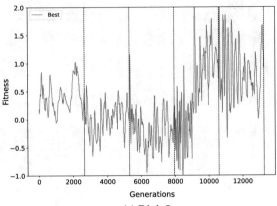

(a) Digit 0.

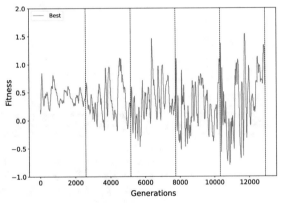

(b) Digit 4.

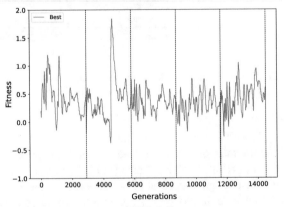

(c) Digit 7.

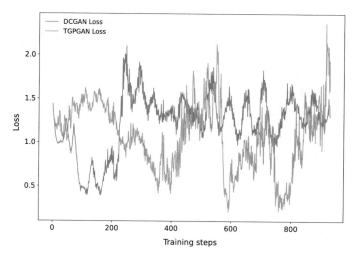

Fig. 8 Loss values for the Discriminator networks of both models across training steps for digit 0

of both models throughout the epochs. We can observe the oscillating behavior across the training steps in both cases. The amplitude of changes appears to be more significant for TGPGAN, but at the time of writing, the significance of this fact is unclear. Many factors can also explain it (e.g., the nature of the images created by the Generators of each model is dramatically different, this alone can be enough to explain these apparent differences). The results appear to indicate that TGPGAN can generate images that the Discriminator has difficulty learning to classify as fake due to the persistent high peaks that surpass the ones of DCGAN. Reaching high values of loss may also be an indicator that the evolutionary process of the TGPGAN is generating samples that follow the same distribution of the training dataset, but that the Discriminator never "saw". This interpretation appears to be confirmed by the comparison of the visual outputs of both TGPGAN and DCGAN, which we present next.

As can be observed by perusing Fig. 10, which presents a sample of the fittest instances created by DCGAN for each digit, some of images are underwhelming. In particular, the images that correspond to digits 3, 4, and 5 are not suitable examples of the corresponding class. Additionally, most of the samples include noisy backgrounds. This contrasts with the TGPGAN images presented in Fig. 9 which appear to be more refined, presenting clear backgrounds and well-defined digits. The combination of these results indicates that in the considered experimental settings: *(i)* performing more training steps would be beneficial to DCGAN; *(ii)* TGPGAN requires less training steps than DCGAN; *(iii)* the examples generated by TGPGAN throughout the process were more valuable to learn the desired classes than the ones created by DCGAN. In addition to generating samples that follow the distribution of the dataset, TGPGAN was also able to generate images that are uncommon or

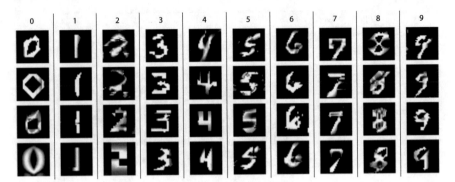

Fig. 9 Random sample of the fittest individuals stored in TGPGAN's archive throughout the training process. Each column corresponds to a given digit, from 0 (leftmost column) up to 9 (rightmost column)

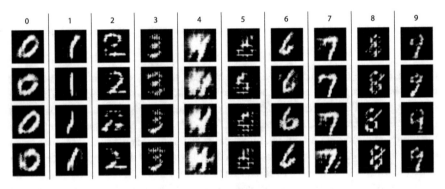

Fig. 10 Random sample of the fittest individuals generated throughout the training process for the DCGAN. Each column corresponds to a given digit, from 0 (leftmost column) up to 9 (rightmost column)

deemed as new under the context of the training dataset, e.g., the last row of digits 2 and 0, the first row of digit 8, and the third row of digit 4.

A final word goes to the computational cost of running DCGAN and TGPGAN. The question is how this comparison should be performed. For instance, performing the same number of training steps for both approaches is probably not adequate since the quality of the results is not similar. Moreover, the performance of TGPGAN depends on parameters such as tree depth, function set, population size, and the number of generations. None of these parameters was optimized for speed; here we are mainly concerned with assessing the approach's feasibility. Nevertheless, although we are comparing apples and oranges, it is necessary to understand the magnitude of the differences. Roughly speaking, using an NVIDIA 3080 GPU, DCGAN and TGPGAN took, respectively, 27 and 5280s (approximately). As such, TGPGAN is approximately 195 times slower than DCGAN. This results from two factors: (i) TGPGAN generates 14.39 times more images, rendering each image 13.59 times slower on average. The number of images necessary to reach the results by starting with a lower number of generations per training step and by reducing the increment of generations per training step. Speeding up the time necessary to evaluate each image can be accomplished by reducing the maximum tree depth, which appears to be excessive, by eliminating costly functions, and by further improving TensorGP. Independently of these possible improvements, it is also necessary to develop means to allow the fair comparison of adversarial models.

7 Conclusions and Future Work

We begin by demonstrating the limitations of ANN-guided evolutionary approaches by performing a set of experiments using a GP engine to generate images that are classified as digits by a state-of-the-art classifier trained on the MNIST dataset. Although the classifier attains a test accuracy of 99.3% when confronted with MNIST instances, it is easily fooled by the evolutionary algorithm that explores and exploits its deficiencies.

To overcome these limitations, we introduce TGPGAN, which can be seen as a variation of the standard DCGAN model, which replaces the typical CNN generator with a GP approach.

The experimental results demonstrate the advantages of the adversarial model over the previous one, allowing the evolution of compelling images that can be correctly classified as digits and Discriminators that are not easily fooled by trivial expression-based GP images.

Comparing the results of TGPGAN with a standard DCGAN is somewhat subjective. Nevertheless, it is safe to say that, for the task at hand, TGPGAN is *at least* competitive with DCGAN, being able to generate high-quality images in a lower number of training steps. The downfall of the approach is the computational cost of GP. To alleviate this burden, we resort to TensorGP, a GP that uses TensorFlow to take advantage of GPU acceleration.

Using expression-based GP has some inherent advantages: *(i)* there is no need to use a priori knowledge for image generation; *(ii)* GP is less likely to match precisely the training data distribution, allowing the generation of valid samples that deviate from the dataset distribution; *(iii)* expression-based images are resolution-independent, allowing the creation of large-format images.

Further work will focus on the application of TGPGAN to other datasets, on the exploration of more advanced archive management mechanisms, and on the comparison of the quality of the Discriminators and Generators produced by TGPGAN and conventional DCGAN. Likewise, testing the performance of both approaches to deal with adversarial hardened Discriminators and to generate such discriminators is also a valuable research direction.

Acknowledgements This work is funded by national funds through the FCT—Foundation for Science and Technology, I.P./MCTES through national funds (PIDDAC), within the scope of

CISUC R&D Unit—UIDB/00326/2020 or project code UIDP/00326/2020 and by European Social Fund, through the Regional Operational Program Centro 2020, under the FCT grant SFRH/BD/08254/2021.

References

1. Arjovsky, M., Chintala, S., Bottou, L.: Wasserstein generative adversarial networks. In: International Conference on Machine Learning, pp. 214–223. PMLR (2017)
2. Baeta, F., Correia, J., Martins, T., Machado, P.: Speed benchmarking of genetic programming frameworks. In: GECCO, pp. 768–775. ACM (2021)
3. Baeta, F., Correia, J., Martins, T., Machado, P.: TensorGP - genetic programming engine in TensorFlow. In: Applications of Evolutionary Computation - 24th International Conference, Evo Applications 2021. pp, 763–778. Springer (2021)
4. Baluja, S., Pomerleau, D., Jochem, T.: Towards automated artificial evolution for computer-generated images. Connect. Sci. **6**(2–3), 325–354 (1994)
5. Cayton, L.: Algorithms for manifold learning. University of California at San Diego Technical Report **12**(1–17), 1 (2005)
6. Correia, J., Machado, P., Romero, J., Carballal, A.: Evolving figurative images using expression-based evolutionary art. In: Proceedings of the fourth International Conference on Computational Creativity (ICCC), pp. 24–31 (2013)
7. Correia, J., Machado, P., Romero, J., Martins, P., Cardoso, F.A.: Breaking the mould an evolutionary quest for innovation through style change. In: Computational Creativity, pp. 353–398. Springer (2019)
8. Correia, J., Martins, T., Machado, P.: Evolutionary data augmentation in deep face detection. In: GECCO (Companion), pp. 163–164. ACM (2019)
9. Correia, J., Martins, T., Martins, P., Machado, P.: X-faces: the eXploit is out there. In: ICCC, pp. 164–171. Sony CSL Paris, France (2016)
10. Costa, V., Lourenço, N., Correia, J., Machado, P.: COEGAN: evaluating the coevolution effect in generative adversarial networks. In: GECCO, pp. 374–382. ACM (2019)
11. Costa, V., Lourenço, N., Correia, J., Machado, P.: Demonstrating the evolution of GANs through t-SNE. In: EvoApplications. Lecture Notes in Computer Science, vol. 12694, pp. 618–633. Springer (2021)
12. Deng, L.: The MNIST database of handwritten digit images for machine learning research. IEEE Signal Process. Mag. **29**(6), 141–142 (2012)
13. Ekern, E.G., Gambäck, B.: Interactive, efficient and creative image generation using compositional pattern-producing networks. In: EvoMUSART. Lecture Notes in Computer Science, vol. 12693, pp. 131–146. Springer (2021)
14. Goodfellow, I.: NIPS 2016 tutorial: generative adversarial networks (2017). https://doi.org/10.48550/ARXIV.1701.00160
15. Goodfellow, I., Pouget-Abadie, J., Mirza, M., Xu, B., Warde-Farley, D., Ozair, S., Courville, A., Bengio, Y.: Generative adversarial nets. Adv. Neural Inf. Process. Syst. **27** (2014)
16. Greenfield, G.R.: Evolving expressions and art by choice. Leonardo **33**(2), 93–99 (2000)
17. Greenfield, G.R.: Tilings of sequences of co-evolved images. In: Raidl, G.R., Cagnoni, S., Branke, J., Corne, D., Drechsler, R., Jin, Y., Johnson, C.G., Machado, P., Marchiori, E., Rothlauf, F., Smith, G.D., Squillero, G. (eds.) Applications of Evolutionary Computing, EvoWorkshops 2004: EvoBIO, EvoCOMNET, EvoHOT, EvoIASP, EvoMUSART, and EvoSTOC, Coimbra, Portugal, April 5–7, 2004, Proceedings. Lecture Notes in Computer Science, vol. 3005, pp. 427–436. Springer (2004). https://doi.org/10.1007/978-3-540-24653-4_44
18. Ha, D.: The frog of CIFAR 10. blog.otoro.net (2016). https://blog.otoro.net/2016/04/06/the-frog-of-cifar-10/

19. Ha, D.: Generating large images from latent vectors. blog.otoro.net (2016). https://blog.otoro.net/2016/04/01/generating-large-images-from-latent-vectors/
20. Hemberg, E., Toutouh, J., Al-Dujaili, A., Schmiedlechner, T., O'Reilly, U.M.: Spatial coevolution for generative adversarial network training. ACM Trans. Evol. Learn. Optim. **1**(2) (2021). https://doi.org/10.1145/3458845
21. Koza, J.R.: Genetic Programming: On the Programming of Computers by Means of Natural Selection. MIT Press, Cambridge (1992)
22. Lewis, M.: Evolutionary visual art and design. In: Romero, J., Machado, P. (eds.) The Art of Artificial Evolution: A Handbook on Evolutionary Art And Music, pp. 3–37. Springer (2008)
23. Machado, P., Cardoso, A.: NEvAr - the assessment of an evolutionary art tool. In: Wiggins, G. (ed.) AISB'00 Symposium on Creative and Cultural Aspects and Applications of AI and Cognitive Science. Birmingham, UK (2000)
24. Machado, P., Cardoso, A.: All the truth about NEvAr. Appl. Intell. **16**(2), 101–118 (2002)
25. Machado, P., Correia, J., Romero, J.: Expression-based evolution of faces. In: Evolutionary and Biologically Inspired Music, Sound, Art and Design - First International Conference, Evo-MUSART 2012, Málaga, Spain, April 11–13, 2012. Proceedings. Lecture Notes in Computer Science, vol. 7247, pp. 187–198. Springer (2012). https://doi.org/10.1007/978-3-642-29142-5_17
26. Machado, P., Correia, J., Romero, J.: Improving face detection. In: Moraglio, A., Silva, S., Krawiec, K., Machado, P., Cotta, C. (eds.) Genetic Programming - 15th European Conference, EuroGP 2012, Málaga, Spain, April 11–13, 2012. Proceedings. Lecture Notes in Computer Science, vol. 7244, pp. 73–84. Springer (2012). https://doi.org/10.1007/978-3-642-29139-5_7
27. Machado, P., Romero, J., Greenfield, G.: Artificial Intelligence and the Arts - Computational Creativity, Artistic Behavior, and Tools for Creatives. Springer (2021). https://doi.org/10.1007/978-3-030-59475-6
28. Machado, P., Romero, J., Manaris, B.: Experiments in computational aesthetics: An iterative approach to stylistic change in evolutionary art. In: Romero, J., Machado, P. (eds.) The Art of Artificial Evolution: A Handbook on Evolutionary Art and Music, pp. 381–415. Springer, Berlin (2007). https://doi.org/10.1007/978-3-540-72877-1_18
29. Machado, P., Vinhas, A., Correia, J., Ekárt, A.: Evolving ambiguous images. In: Yang, Q., Wooldridge, M. (eds.) Proceedings of the Twenty-Fourth International Joint Conference on Artificial Intelligence, IJCAI 2015, Buenos Aires, Argentina, July 25–31, 2015, pp. 2473–2479. AAAI Press (2015). http://ijcai.org/papers15/Abstracts/IJCAI15-350.html
30. Martins, T., Correia, J., Costa, E., Machado, P.: Evolving stencils for typefaces: combining machine learning, user's preferences and novelty. Complexity **2019** (2019). https://doi.org/10.1155/2019/3509263
31. Metz, L., Gulrajani, I.: Compositional pattern producing GAN. In: NeurIPS Workshops, vol. 1 (2017)
32. Nguyen, A.M., Yosinski, J., Clune, J.: Deep neural networks are easily fooled: High confidence predictions for unrecognizable images. In: IEEE Conference on Computer Vision and Pattern Recognition, CVPR 2015, Boston, MA, USA, June 7–12, 2015, pp. 427–436. IEEE Computer Society (2015)
33. Romero, J., Machado, P. (eds.): The Art of Artificial Evolution: A Handbook on Evolutionary Art and Music. Natural Computing Series. Springer (2008)
34. Rooke, S.: Eons of genetically evolved algorithmic images. In: Bentley, P.J., Corne, D.W. (eds.) Creative Evolutionary Systems, pp. 339–365. Morgan Kaufmann (2002)
35. Secretan, J., Beato, N.: Picbreeder: evolving pictures collaboratively online. In: Czerwinski, M., Lund, A.M., Tan, D.S. (eds.) Proceedings of the 2008 Conference on Human Factors in Computing Systems, CHI 2008, 2008, Florence, Italy, April 5–10, 2008, pp. 1759–1768. ACM (2008). https://doi.org/10.1145/1357054.1357328
36. Secretan, J., Beato, N., D'Ambrosio, D.B., Rodriguez, A., Campbell, A., Folsom-Kovarik, J.T., Stanley, K.O.: Picbreeder: a case study in collaborative evolutionary exploration of design space. Evol. Comput. **19**(3), 373–403 (2011)

37. Sims, K.: Artificial evolution for computer graphics. In: SIGGRAPH '91: Proceedings of the 18th Annual Conference on Computer Graphics and Interactive Techniques, pp. 319–328. ACM, New York (1991). https://doi.org/10.1145/122718.122752
38. Spector, L., Alpern, A.: Criticism, culture and the automatic generation of artworks. In: Proceedings of the Twelfth National Conference on Artificial Intelligence, pp. 3–8. AAAI Press, Menlo Park (1994)
39. Staats, K., Pantridge, E., Cavaglia, M., Milovanov, I., Aniyan, A.: TensorFlow enabled genetic programming. In: Proceedings of the Genetic and Evolutionary Computation Conference Companion. pp. 1872–1879. ACM (2017)
40. Stanley, K.O.: Compositional pattern producing networks: a novel abstraction of development. Genet. Program Evolvable Mach. **8**(2), 131–162 (2007)
41. Stanley, K.O., Miikkulainen, R.: Evolving neural networks through augmenting topologies. Evol. Comput. **10**(2), 99–127 (2002)
42. Todd, S., Latham, W.: Evolutionary Art and Computers. Academic (1992)
43. Unemi, T.: SBART 2.4: breeding 2D CG images and movies and creating a type of collage. In: Knowledge-Based Intelligent Information Engineering Systems, 1999. Third International Conference, pp. 288–291. IEEE (1999)
44. Wang, C., Xu, C., Yao, X., Tao, D.: Evolutionary generative adversarial networks. IEEE Trans. Evol. Comput. **23**(6), 921–934 (2019)
45. Xiao, H., Rasul, K., Vollgraf, R.: Fashion-MNIST: a novel image dataset for benchmarking machine learning algorithms (2017). arXiv preprint arXiv:1708.07747
46. Zhu, J.Y., Park, T., Isola, P., Efros, A.A.: Unpaired image-to-image translation using cycle-consistent adversarial networks. In: Proceedings of the IEEE International Conference on Computer Vision, pp. 2223–2232 (2017)

Modeling Hierarchical Architectures with Genetic Programming and Neuroscience Knowledge for Image Classification Through Inferential Knowledge

Gustavo Olague, Matthieu Olague, Gerardo Ibarra-Vazquez,
Isnardo Reducindo, Aaron Barrera, Axel Martinez, and Jose Luis Briseño

Abstract Brain programming is a methodology based on the idea that templates are necessary to describe artificial dorsal and ventral streams and their combination into an artificial visual cortex. We review the main concerns by introducing some initial thoughts about the status of genetic programming and other methodologies related to our research work. This chapter proposes the hierarchical integration of two architectures (templates) to enhance the quality of acquiring artificial visual percepts. We theoretically justified the necessity for designing manual hierarchical architectures. Planning complex structures through inferential knowledge simplify the design while adopting current technology. The methodology base its analysis on providing domain knowledge (neuroscience) at a higher level while looking for better computational structures within a local (lower) level. The efficiency of searching for optimal

G. Olague (✉) · A. Barrera · A. Martinez · J. L. Briseño
EvoVisión Laboratory, CICESE, Ensenada, B.C. 22870, Mexico
e-mail: olague@cicese.mx

A. Barrera
e-mail: abarrera@cicese.edu.mx

A. Martinez
e-mail: amartinez@cicese.edu.mx

J. L. Briseño
e-mail: briseno@cicese.mx

M. Olague
Anahuac University Queretaro, 76246 El Márques, Querétaro, Mexico
e-mail: matthieu.olague03@anahuac.mx

G. Ibarra-Vazquez
ITESM, Institute for Future of Education, Monterrey, N.L., 64849, Mexico

I. Reducindo
Autonomous University of San Luis Potosí, Information Sciences Faculty, Fracc. Talleres, San Luis Potosí 78494, Mexico
e-mail: isnardo.reducindo@uaslp.mx

© The Author(s), under exclusive license to Springer Nature Singapore Pte Ltd. 2023
L. Trujillo et al. (eds.), *Genetic Programming Theory and Practice XIX*,
Genetic and Evolutionary Computation, https://doi.org/10.1007/978-981-19-8460-0_7

architectural configurations proceeds from deductive and inductive reasoning. This chapter brings a proposal of abductive reasoning to enrich the brain programming paradigm by taking advantage of computational re-use of dorsal stream discoveries while enhancing the overall complexity of the final proposal. We propose a Visual Turing test to establish the quality of the proposal in comparison with the state of the art. The results show that our methodology can produce consistent outcomes during training and testing, representing significant progress toward thought representation.

1 Introduction

This chapter studies artificial intelligence (AI) from symbolic learning, considering image classification problems. Symbolic learning is one of the classical approaches to AI, whose development has been complex regarding image classification due in part to the lack of a theory to create sophisticated representations. Symbolic learning is associated with identifying mathematical expressions trained with sets of images. The study of symbolic learning for image classification through genetic programming (GP) framework currently lacks a transparent methodology able to beat other state-of-the-art approaches [11].

This document introduces the idea of abductive reasoning in GP considering hierarchical structures inspired by the human visual cortex. We study the problem of symbolic representation under image classification following a variant of the Turing test, which is a method of inquiry in AI for determining or not if a computer is capable of thinking. The original test does not require that the machine intelligence replicates thinking like a human being, only that a computer could trick a human into believing that the machine is human. Nonetheless, the seminal test requires the device to exhibit some understanding of natural language processing (NLP). In our case, we apply a visual Turing test based only on identifying visual information unrelated to language use but on understanding visual concepts, i.e., the identification and association of visual stimulus with emotion recognition.

To outline an approach departing from current machine learning (ML), we took ideas from inferential knowledge; see Sect. 3. Our strategy identifies different types of reasoning susceptible to being transcribed into the form of computer algorithms. The methodology follows conceptual representation in deductive, inductive, abductive, retroductive, and transductive reasoning. With these elements, we introduce the first proposal of abductive logic based on the results of previous research through the combination of deductive and inductive methodologies. The association of earlier studies regarding salient object detection and image classification influences the proposal's value since the new plan significantly increases the size of solutions while helping to find answers to more complex problems and taking advantage of the program searches made in previous studies.

Nowadays, the size of computer solutions to computer vision (CV) problems like image classification reaches unique requirements that digital companies can only achieve many times. For example, Google trained in 2021 a two billion parameter

AI vision model on three billion images while achieving 90.45% top-1 accuracy on ImageNet. Also, in 2021, Facebook implemented a SElf-supERvised (SEER) approach for CV inspired by NLP with billion parameters to learn models from any group of images on the internet. Indeed, regarding NLP, the Generative Pre-trained Transformer 3 (GPT-3) has 175 billion parameters, and basically, the architecture design of all these networks follows a handcrafted approach. This situation makes us think about the unsuitability of searching for optimal architecture design through genetic and evolutionary computation (EC) approaches, at least with currently available technology.

We pause our discussion in Sect. 4.1 to a theoretical biological problem related to the Word Links game to illustrate the problem of blindly attempting to search for a solution from scratch. The frequency and distribution of functional amino acid sequences result in colossal search spaces, making it impossible to automate the search. This situation helps us to introduce our approach to designing symbolic representations through what we call templates inspired by classical neuroscientific representations that are not based on neural networks but rest on mathematical modeling. It is possible to extrapolate the template idea to neural architecture search.

Moreover, as we will explain here, the images have the constraint that the learned concepts must be contingent. Programming a computer to represent a contingent thought implies that the concept may be true or false, whereas non-contingent thoughts are necessarily true or false. In other words, databases portraying concepts like ImageNet contain empirical knowledge, while databases characterized by psychological information like OASIS include information that can be known through reason, by analysis of concepts, and by valid inferences.

GP is a technique for automatic programming (AP) centered on the problem of machine learning. The idea is to automatically build computer programs to produce some machine intelligence by approximately solving challenging computational tasks. In this way, our research touches on several areas: automatic programming, machine learning, computer vision, artificial intelligence, and evolutionary computation.

Exploring the opportunities and limitations of current genetic programming and related research areas help in understanding the reasons behind past failures and future success. Hence, we propose to review the subject since it has roots in several myths and open issues around automatic programming, genetic programming, and mainstream research on computer vision and deep learning (DL).

2 Myths and Prospects of Genetic Programming

Scientists and Engineers understand automatic programming as computer programming where machines generate code under certain specifications. Translators could be considered automatic programs mapping high-level language to lower-level language, i.e., compilers. Automatic programming includes applying standard libraries into what is known as generative programming, so the programmer does not need to

Table 1 Assessment of myths on automatic programming in other research areas

Myth	Method		
	Machine learning	Computer vision	Evolutionary computation
(a)	✓	✗	✓
(b)	✓	✗	✓
(c)	✗	✗	✗

re-implement or even know how some piece of code works. Also, the idea of creating source code based on templates or models based on a visually graphic interface where the programmer is instead a designer who, through drag and drop functions, defines how the app works without ever typing any lines of code. In 1988, Rich and Waters identified some myths and realities about automatic programming, which are still relevant today [23]. Next, we present a list of the relevant points helpful to our exposition:

(a) The myth implies that automatic programming does not need domain knowledge.
(b) The myth is that general purpose and fully automatic programming are possible.
(c) The myth indicates that there will be no more programming.

We could extrapolate these three myths into ML, CV, and EC to give a snapshot of their actuality (Table 1). Regarding ML, the myths of unnecessary domain knowledge, general purpose, and fully automatic programming are at the methodology's core, while the third myth is not a pursued property. For CV, none of the myths apply since this discipline centers on the study of vision to recreate visual perception on a problem-by-problem basis. In the case of EC, we found the same situation concerning ML. Nevertheless, genetic programming as an approach aiming at AP, we consider that all myths apply in general.

GP is a research paradigm driven by the idea "tell the machine what to do, not how to do it", as explained by Koza [12] while paraphrasing the words of Samuel [27]. The methodology centers on the goal of accomplishing program induction as the process for building complete working programs. Nevertheless, the solutions miss the property of scalability. A pervasive myth anchored on the idea that if we want to achieve complete automatic programming, GP requires only general principles and no human assistance in such a way of being domain independent. Paradigmatically, the kind of solutions is usually narrow and specific [13]. Indeed, Koza predicted in 2010 that GP would enhance due to Moore's law or the increased availability of computer power—a situation that did not arrive. This chapter presents the idea that to manage an ever-increasing problem; we need a way to enrich the problem representation paired with a scheme that processes inferential knowledge. O'Neill and Spector equate the challenge of achieving automatic programming with reaching a complete solution to the AI problem [22]. We propose a novel visual Turing test as an instance of an unsolved problem that can drive us in the long term to study the scheme introduced in this chapter.

We believe that many authors recognize the issues and implications of follow-ing the myths and hide part of their approach since this will be problematic at the moment of submitting their work for peer review or simply because they are so involved with the paradigm that they do not stop to report the other methods used in the investigation. Also, many people working on GP study the methodology as part of ML without looking into AP, while the opposite is rare. We believe that GP encompasses both research areas and considers it when developing solutions to CV problems. Indeed, it is unrealistic to expect that GP alone will be sufficient to achieve automatic programming, and researchers usually incorporate other methods without correctly reporting them in the manuscripts. We do not claim that the scheme reported here is novel and left other researchers the opportunity to clarify how they approached their work according to the ideas presented in this chapter. Neverthe-less, the innovative representation follows mainstream cognitive science combined with GP to create adaptatively complex models of visual streams to approach salient object detection and image classification problems. Also, our contribution roots in the scheme application to architectural representations and how to increase solutions complexity through the hierarchical combination of architectures.

In 2010, O'Neill et al. recognized several open issues that people working on GP need to address if further development of the area is achievable in such a way to realize the full potential and become conventional in the computational problem-solving toolkit [21]. Next, we recall the main open issues relevant to our discussion:

(a) GP representation, modularity, complexity, and scalability. The topic relates to identifying appropriate (optimal) representations for GP while considering struc-ture adaptation based on some measure of quality that captures the relationship between the fitness landscape and the search process. Also, the idea relates to the tenet that the invention of accompanying modularity and complexity of function at a high level originates from the development or growth of simple building blocks. Regarding scalability or the ability to provide algorithmic solutions to problems of substantial size/dimensionality, the authors recognize the difficulty since the GP process always tends to compromise between the goals set by the fitness function and the program's complexity. Authors claim that GP's strength resides in providing a natural engine for generalization, whose obstacles can be overcome only with recipes for modularity and scalability.

(b) Domain knowledge. This subject refers to the idea of defining an appropriate AI ratio. Koza highlighted that for a minimum amount of domain knowledge supplied by the user (the intelligence), GP achieves a high return (the artificial). All this situation is about attaining human-competitive results automatically and partly measured by the existence of a high AI ratio.

(c) GP benchmarks and problem difficulty. This issue relates to the problem of determining a set of test problems that the scientific community can rigorously evaluate, and as a result, the algorithmic discoveries can be accepted based on such benchmarks.

(d) GP generalization, robustness, and code re-use. These elements refer to the nature of program representation and the qualities the designer attempts to fulfill. As stated by the authors, GP generalization relates to ML and the idea of overfitting as in statistical analysis; therefore, such concepts relate to the robustness of solutions and code re-use.

According to the authors listing the above open issues, GP was not universally recognized as mainstream and trusted problem-solving strategy despite numerous examples where the technique outperforms other ML methods. They argue that the resistance of the ML community to embrace GP is rooted in Darwinian thinking. Instead, we believe it is more related to the overuse of random processes and the implicit connection taken for granted that structures raise from fitness. The second statement is made explicit by Koza in such a way that the GP paradigm address the problem of getting computers to learn to program themselves by providing a domain-independent way to search the space of possible computer programs for a program that solves a given problem. This idea resides on the derivation of knowledge following a way to do program induction. In the words of Koza:

> "...That is, genetic programming can search the space of possible computer programs for an individual computer program that is highly fit in solving (or approximately solving) the problem at hand. The computer program (i.e., structure) that emerges from the genetic programming paradigm is a consequence of fitness. That is fitness begets the needed program structure." (Koza, 1992, p. 3)

After carefully reading the O'Neill et al. article, we assume that such principle anchors in the imaginary and that designers believe it is true-valid when designing a search procedure for the artificial induction of programs. As we will propose in this chapter, we believe that to achieve program synthesis; we need a way of replicating the theory of inferential knowledge in the sense that to advance the search for complex programs when attempting to solve challenging problems, we need to recognize not only the idea of induction, but also other ways of inferential knowledge like deductive reasoning. The whole scientific approach requires a complete cycle of deductive–inductive reasoning. We understand that this idea of structure raising from fitness is a myth like those identified in automatic programming.

Again we can extrapolate the identified open issues to observe their relevance in the proposed three research areas (Table 2). Representation, modularity, complexity, and scalability are open issues in ML, CV, and EC. On the other hand, generalization, robustness, and code re-use are well-studied subjects with clear theoretical and practical guides in all three research areas. Regarding domain knowledge, ML and CV incorporate it systematically as part of their approach to problem-solving; nonetheless, EC follows a similar path of leaving the charge of building intelligent solutions from apparently disconnected building blocks based on the search process. Finally, benchmarks and problem difficulty are pervasive problems in ML and EC since both attempt to discover algorithms that can be useful in an extensive range of

Table 2 Assessment of GP—open issues in other research areas

Myth	Method		
	Machine learning	Computer vision	Evolutionary computation
(a)	✓	✓	✓
(b)	✗	✗	✓
(c)	✓	✗	✓
(d)	✗	✗	✗

different problems, which is not the case in CV, where each problem follows strict formulations.

Nowadays, the situation is still more worrisome with the worldwide impact of DL since this technology is not even contrasted with alternative formulations because most people in computer science estimate that there is no competitive alternative. However, such a paradigm has numerous sometimes deadly failures [3], and since overcoming such hurdles is not an easy task, the always present possibility of lack of success represents an opportunity for other research paradigms [11]. Here, we recall some aspects where neural networks' lack of success impact trust and confidence in these systems:

(a) Brittleness and lack of trustworthiness. These aspects refer to the lack of invariance whose attempts to overcome lay on extending the number of patterns, including rotation, scaling, illumination, and as many as different representations of a kind of object. Nevertheless, input data corruption through adversarial attacks is a severe source of difficulty. Moreover, contamination with information in the form of stickers or other ways of marking traffic signals, cars, persons, or other objects represents real conundrums for such technology.

(b) Uncertainty. Classification systems based on DL report high certainty for the training data, but the way of calculating robustness is still an open issue. Data-dependent methodologies need improvement when calculating and dealing with uncertainty toward a robust approach whose accuracy refines its confidence in the outcome.

(c) Data-driven technology and expensive hardware requirements. One of the most significant disadvantages of deep learning is that it requires considerable data to perform better than other techniques. It is costly to train due to complex data models. Thus, this technology requires expensive Graphic Processing Units (GPUs) and hundreds of machines. This data dependency is the source of most of the problems described in this section.

(d) Embedded bias and catastrophic forgetting. Bias is the Achilles heel of data analysis, divided into data bias and bias in evaluating or creating data. DL as ML depends on accurate, clean, and well-labeled training data to learn from so they can produce accurate solutions. Most AI projects rely on data collection, cleaning, preparation, and labeling steps, which are sources of bias or catastrophic

Table 3 Assessment of DL–lack of success in other research areas

Myth	Method		
	Machine learning	Computer vision	Evolutionary computation
(a)	✗	✗	✓
(b)	✗	✗	✓
(c)	✓	✗	✓
(d)	✗	✗	✗
(e)	✗	✗	✓

forgetting. This last issue results from updating data, like in the case of detecting artificially generated fake images. In order to repair the performance of CNNs, the researchers trained them with new data, and as the cycle continued, the CNN forgot how to detect the old ones, even the original clean data.

(e) Explainability. The advantage of an end-to-end methodology combined with the black-box paradigm makes it hard to conclude why the system selects certain features, especially for problems with many classes or when different networks provide solutions to the same problem. How DL reaches conclusions is a mysterious process that is somewhat stuck on the theoretical side and left to designing better, more constrained datasets.

Table 3 summarizes the analysis of these issues in the other three research areas that can complete our critique of the situation in optimization and learning approaches. Brittleness and lack of trustworthiness are not issues for ML and CV, but these are problems for EC. Researchers working in ML and CV do not consider EC as part of their popular methodologies; therefore, it is necessary to work on these aspects to achieve widespread acceptability. Uncertainty and explainability are also red flags for EC, while both are well-studied in ML and CV. In the case of data-driven technology and expensive hardware requirements, ML and EC have issues, while CV has no problems. Finally, none of the other research areas has bias and catastrophic forgetting issues.

Next, we propose to apply some strategies founded on inferential logic for building knowledge to enhance designed models to approach more challenging problems. Thus, we start to explore the scheme proposed in this chapter as a way to create complex solutions based on hierarchical architectures of the visual stream.

3　Inferential Knowledge

There are four types of knowledge representation in AI: relational knowledge, inheritable knowledge, inferential knowledge, and procedural knowledge [14]. The first refers to storing facts, and it founds application in relational database systems. The second deals with ways to represent data with some hierarchy, subclasses inherited from the superclasses in the representation—the superclass holds all the data in the subclasses. The third is essential for our discussion since inferential knowledge defines understanding in terms of formal logic conditions and has a strict rule. The knowledge raises from objects by studying their relation. Finally, procedural knowledge in AI represents control information in small programs and codes to describe how to proceed and do specific tasks, i.e., usually if-then rules. Paradigmatically, this is what we need to answer Samuel's question, but first, we need an approach to create knowledge.

Knowledge representation needs a set of strict rules which can derive more facts, verify new statements, and ensure correctness. Many inference procedures are available from several types of reasoning. The reasoning is part of intelligence, and to derive artificial ways of creating knowledge, we first need to understand how humans abstract it. Even though humans use several ways of creating knowledge, let us focus on logic first. In logic, inference refers to a process of deriving logical conclusions from premises known or assumed to be true.

An inference is valid if based upon sound evidence and the conclusions follow logically from the premises [16]. The reasoning divides into deductive, inductive, analogical, abductive, cause-and-effect, decompositional, and critical thinking. The cycle of deductive–inductive reasoning/theory as scientific explanation drives the overall conception of science. Theoretical systems like Newton's laws and Kepler's laws show us how to interpret scientific theory in terms of explanations. Examples of modeling visual information to accurately predict complex corners and retro-reflective targets in terms of morphology, geometry, and physics is made with regression following deductive reasoning despite application of optimization techniques [17]. Next, we focus on a typical division made of the following five types of reasoning.

3.1 Deductive Reasoning

This kind of reasoning helps make predictions, which contributes to knowledge. The reasoning starts with ideas to generate hypotheses to produce observations.

$$\text{Testing the Theory} \rightarrow \text{Observations/Findings}$$

The hypothesis is a supposition, preposition, or principle that is supposed or taken for granted to draw a conclusion or inference for proof of the point in question. For deductive reasoning to work, hypotheses must be correct since the conclusion follows with certainty from the premises. Thus, the hypotheses follow examinations to derive logical conclusions—this is true for all members. Therefore, we achieve a theoretical explanation of what we have observed, which is the contribution to knowledge.

Start with Theory

⇓

Derive Hypothesis

⇓

Collect Data

⇓

Analyse Data

⇓

Confirm or Reject Hypothesis

⇓

Revise Theory

This is summarized through the following relationship:

$$\underbrace{\mathscr{A}}_{\text{(Rule)}} + \underbrace{\mathscr{B}}_{\text{(Case)}} = \underbrace{\mathscr{C}}_{\text{(Result)}} \tag{1}$$

and is exemplified with the following syllogism:

(Deductive Reasoning)   All men are mortal;
                        but Seth is a Man.
                        ―――――――――――――――
                        Therefore Seth is a mortal.   ∴

This way of approaching problems is classical of CV and is the method we use in EvoVisión as explained in [17]. Nevertheless, the following method is classical to GP, and this chapter attempts to expose the differences and complementaries to adopt more complex methodologies for problem-solving.

3.2 Inductive Reasoning

Inductive research begins with a research question and empirical data collected to generate a hypothesis and theory.

Observations/Findings → Testing the Theory

It is the inverse of the deductive approach. Inductive learning starts with the phenomena, and the researcher needs to be careful about the inferred rules since these are based on observations, not speculations. The observations produce patterns, and we obtain rules or theories from these. Therefore, the conclusion follows not with certainty but only with some probability.

Observe and Collect Data

⇓

Analyse Data

⇓

Look for Patterns in Data

⇓

Develop Theory

This is summarized through the following relationship:

$$\underbrace{\mathscr{A}}_{\text{(Result)}} + \underbrace{\mathscr{B}}_{\text{(Case)}} = \underbrace{\mathscr{C}}_{\text{(Rule)}} \tag{2}$$

and is exemplified with the following syllogism:

(Inductive Reasoning)
Seth is a mortal;
but Seth is a Man.
_____
Therefore All men are mortal.   ∴

nevertheless, the following syllogism represents an unsound argument

(Inductive Reasoning)
Seth eats meat;
but Seth is a Man.
_____
Therefore All men eat meat.   ∴

The researcher must ensure that all generalizations are in the theory or rules since the contribution to knowledge should be routed in a sound argument.

3.3 Abductive Reasoning

Note the same construct in a different order; therefore, in inferential knowledge, we get different research-design approaches. This process helps derive the contribution to knowledge. In the abductive approach, we must provide the best possible explanation for what we have observed, even if the observation is incomplete. Note that this does not refer to gathering incomplete data and calls it abductive. The best explanation of incomplete observation uses rules plus results. We obtain an explanation from observations (unexpected observation—little surprise) and an idea (rules or theories). This approach is practical for testing hypotheses (similarities with the process of generalization or transfer learning).

This is summarized through the following relationship:

$$\underbrace{\mathscr{A}}_{\text{(Rule)}} + \underbrace{\mathscr{B}}_{\text{(Result)}} = \underbrace{\mathscr{C}}_{\text{(Case)}} \tag{3}$$

and is exemplified with the following syllogism:

All men are mortal;

(Abductive Reasoning)　but Seth is a mortal.
_____

Therefore Seth is a Man.　∴.

3.4　Retroductive Reasoning

The retroductive approach aims to understand why things are the way they are. This is summarized through the following relationship:

$$\underbrace{\mathscr{A}}_{(Result)} + \underbrace{\mathscr{B}}_{(Case)} = \underbrace{\mathscr{C}}_{(Cause)} \tag{4}$$

and is exemplified with the following syllogism:

Seth is a mortal;

(Retroductive Reasoning)　but Seth is dead.
_____

Therefore What killed Seth.　∴.

3.5　Transductive Reasoning

This is summarized through the following relationship:

$$\underbrace{\mathscr{A}}_{(Result)} + \underbrace{\mathscr{B}}_{(Cause)} = \underbrace{\mathscr{C}}_{(Case)} \tag{5}$$

and is exemplified with the following syllogism:

Seth is a mortal;

(Transductive Reasoning)　but What killed Seth?
_____

Therefore Seth is dead.　∴.

4 Inferential Knowledge in Brain Programming

Brain programming (BP) is a research paradigm based on fusing cognitive computational models used as templates with a powerful search mechanism to discover symbolic substructures embedded within the higher graph structure. The first results were reported in 2012 when an artificial ventral stream was proposed to approach an object recognition problem [4]. The task consists of identifying critical features (interest region detection and feature description) of an artificial "what" pathway to simplify the whole information process, see Fig. 1a. The strategy reduces the total

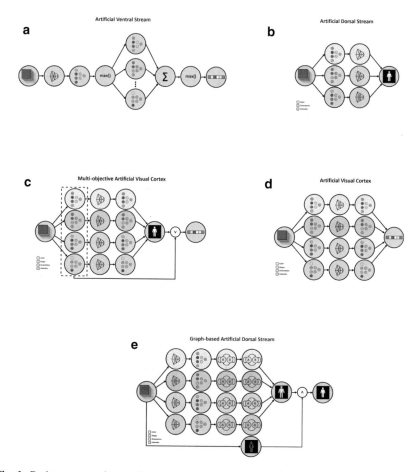

Fig. 1 Brain programming applies a template to characterize the behavior that the designer is attempting to recreate in the algorithm. Here, we show five different algorithms that we reported previously. **a** Object Recognition with AVS [4]. **b** Visual Attention with ADS [6]. **c** Visual Attention plus Object Recognition [9]. **d** Object Recognition with AVC [18]. **e** Graph-based Visual Attention [20]

computational cost by substituting a set of patches with an offline learning process to enforce a functional approach. Later, BP improved an artificial dorsal stream by searching for optimal programs embedded within a visual attention architecture, see Fig. 1b. The process incorporated learning to a handmade technique primarily based on deductive and heuristic reasoning [6].

BP design improved with the idea of studying two tasks simultaneously (salient object detection and image classification) and applying the framework of multi-objective optimization [9]. The design incorporates the V1 stage of visual attention into the artificial ventral stream to create a kind of artificial visual cortex, see Fig. 1c. The design focuses on the computation of the feature descriptor into a particular image region. The results significantly improved the performance of previous ventral stream processing.

An enigmatic result of the artificial visual cortex consists of the random discovery of perfect solutions for non-trivial object recognition tasks [18]. The design incorporates the V1 stage of visual attention and the idea of parallel computation across different visual dimensions, see Fig. 1d. The design considers integrating four visual dimensions after transformations with a set of visual operators discovered by multi-tree GP.

Recently, the EvoVisión laboratory proposed to evolve an ADS following the idea of graph-based visual attention while incorporating the dimension of form and evolving multiple functions with a multi-tree GP [20]. The new algorithm includes an image segmentation process merged with the output of the visual saliency process to create the proto-object, see Fig. 1e.

4.1 Doublets

Lewis Carroll proposed a game called "Word Links" around Christmas 1877 as a form of entertainment for two bored young ladies. The game is a kind of puzzle whose rules Lewis submitted to the journal *Vanity Fair* and published in March 1879.

The pastime consists of proposing two words of the same length, and the puzzle consists in linking these together by interposing other words, each of which shall differ from the next word in one letter only. The puzzle starts with the two words, called doublets,[1] and the player looks for interposing words, called links, by changing one letter from the given words until finishing connecting both words. The entire series, called a chain, from which an example is here:

---

[1] In linguistics-etymology, people call two or more words in the same language doublets or etymological twins or twinlings when they have different phonological forms but the same etymological root.

$$\text{HEAD}$$
$$\text{heal}$$
$$\text{teal}$$
$$\text{tell}$$
$$\text{tall}$$
$$\text{TAIL}$$

What is important to understand is that according to TWL (Tournament Word List) scrabble dictionary, there are 4214 four-letter words with meaning, while there are $26^4 = 456,976$ possible combinations of words from the alphabet. Each and every four-letter word has $25 \times 4 = 100$ neighboring sequences, but only $4214 \times 100/456976 = 0.92$ neighbors, on average, have meaning. This game illustrates a complex system where meaning is not randomly distributed [8].

This popular word game serves as an analogy to understand the problem of frequency and distribution of amino acid sequences which are functional, either as enzymes or in some other way. The alteration of a single letter corresponds to the most straightforward evolutionary step, the substitution of one amino acid for another, and the requirement of meaning corresponds to the requirement that each unit step in evolution should be from one functional protein to another [15]. If natural selection evolves, functional proteins must form a continuous network (series of small isolated islands in a sea of nonsense sequences) that can be traversed by unit mutational steps without passing through non-functional intermediates. Salisbury calculates the search space for a small protein as follows:

A typical small protein contains about 300 amino acids, and its controlling gene about 1,000 nucleotides (three for each amino acid). Because each nucleotide in a chain represents one of four possibilities, the number of different kinds of chains is equal to the number 4 to the power of the number of links in the chain; that is, $4^{1,000}$, or about 10^{600}. (Salisbury 1969, p. 342)

Then, Salisbury imagined a primeval ocean uniformly 2km deep, covering the entire Earth, containing DNA (deoxyribonucleic acid) at an average concentration and each double-stranded molecule with 1,000 nucleotide pairs. Also, he imagined each DNA molecule reproducing itself one million times per second and occurring a single mutation each time a molecule reproduces, while no two DNA molecules are ever alike [26]. Salisbury estimates that in four billion years, the production is about 7.74×10^{64} different kinds of DNA molecules, and if we consider 10^{20} similar planets in the Universe, we obtain a total of 7.74×10^{84} (roughly 10^{85}) different molecules. So he concludes: "If only one DNA molecule were suitable for our act of natural selection, the chances of producing it in these conditions are $10^{85}/10^{600}$ or only 10^{-515}". The size of these numbers receives the name hyper astronomical, and the probability of occurring makes us think of an impossibility.

Regarding brain complexity, researchers base brain organization on regions. Within computer science, it is popular to think of it in terms of neural mechanisms, although other authors explain the evolutionary development based on volume and mass [10]. The human brain contains about 100 billion neurons, more than 100,000 km of interconnections, and has an estimated storage capacity of 1.25×10^{12} bytes. This is equivalent to 10^{15} connections and contains roughly the same number of neurons as there are stars in the Milky Way. The analogy is with an analog device where the explanation of information flow depends on synaptic processing time, conduction speed, pulse width, and neuron density. In other words, the understanding of the brain depends on modeling the information processing capability per unit time of a typical human brain as a function of interconnectivity and axonal conduction speed. Indeed, such an analogy helps create numerical models/methods at the forefront of technological advances. However, a missing part is an analogy with information processing devices that manage language and code like in the doublets example and the DNA. Brain programming proposes to create an analogy with an information process based on symbolic computation. Nowadays, current computational technology is meager compared to small parts of a brain region and understanding how language could be possible within neurons. The idea is to adapt the current (handmade) cognitive computational proposal with GP to incorporate learning. In this chapter, we extended GP beyond purely inductive reasoning while recognizing the need for a deductive approach and the first proposal of abductive learning.

4.2 The Visual Turing Test

Nowadays, thousands of researchers see DL as the holy grail to approach AI since this technique proved that it is possible to solve previously thought challenging problems while ignoring the risks they may pose. As we reviewed in Sect. 1, the GP open issues require a benchmark with a proven problem difficulty that involves domain knowledge and complex representation. The aim is to force the designer on many axes about modularity, scalability, generalization, robustness, and code re-use. Indeed, as a powerful technique, DL moved the limits of AI and launched society into a new stage of computational development. However, as the AI frontier erodes, it is necessary to create new benchmarks that help us to define in practice what we mean by intelligence. The idea of intelligence is not new, and many believe, at least in AI, that Turing adequately defined the term in his now famous test [28]. Nevertheless, the idea has origins that date back over two millennia and are in the Talmud-Sanhedrin 65b:

> Rava created a person, and sent it before Rav Zeira
> Rav Zeira spoke to it, but it would not reply
> Rav Zeira said to it
> you are a creation of one of the magicians, return to your dust

This paragraph is a translation, and there are many versions; however, the key to our exposition is the idea that for an artificial being to be a man, the golem must possess language. The principal method of human communication, consisting of words used in a structured and conventional way and conveyed by speech, writing, or gesture, is a sign of intelligence. The term "intelligent computers" refers to the question "Can computers think" and has ramifications for robotics [24]. There are multiple definitions of AI all around four different axes: (1) thinking-humanly, (2) thinking-rationally, (3) acting-humanly, and (4) acting-rationally [25]. Regarding standard definitions (dictionary), intelligence is the ability to acquire and apply knowledge and skills, the ability to learn or understand or deal with new or trying situations, the skilled use of reason, and the ability to apply knowledge to manipulate one's environment or to think in abstract terms as measured by objective criteria (such as tests). In simple terms, the act of understanding. This last idea developed within the Thomistic tradition was initially expressed in the writings chiefly by St. Thomas Aquinas [2] and related to the psyche (soul) to acquire knowledge through the internal and external senses[2] using reasoning. In other words, humankind's intelligence is:

> Human intelligence is the ability to read our experiences and bring to light what is intelligible within them.

The definition is profound in that we are not required to ask questions about the universe's intelligibility, but we are only required to recognize/understand the world. In other words, something that can be understood by the intellect, not by the external senses. Knowledge is more than sensory experience; but it begins in the senses. If we did not have the sensory experience of the world around us, our minds would be empty.[3] In this way, understanding goes beyond perception, like when sight observes the world and our soul (psyche) moves our (internal senses) emotions, i.e., watching our baby fires deep feelings of love. Thus, the term experience goes beyond empirical knowledge and touches the reality we undergo by reasoning and will; this last one is the power that inclines us to what is apprehended as good or fitting.

Figure 2 provides examples of images requiring cognitive abilities to correctly identify the right emotion despite the similarity in visual patterns. The images are superposed with the circumplex model of attention that psychologists apply using valence and arousal scores obtained from directly asking a person who observes the images to rate them. The problem of correctly identifying the right emotion becomes harder to solve by computational methods, as illustrated in Figs. 3 and 4. In both collages, we observe the diversity of patterns that preclude us from directly solving the identification task, and it makes us think of Wittgenstein's beetle [29]. The beetle

---

[2] According to St. Thomas Aquinas, internal senses are common sense, imagination, estimative power, and memory; and external senses are smell, sight, touch, taste, and hearing.

[3] CV attempts to recreate the external sense of sight that precedes language; hence representation of thought is the goal we strive to reach by formulating a representation helpful to communicating ideas.

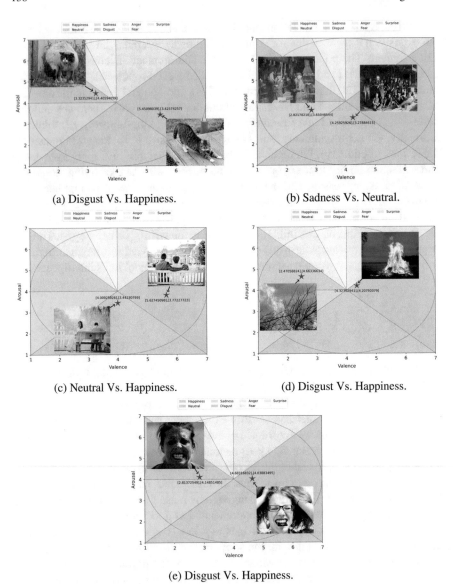

(a) Disgust Vs. Happiness.

(b) Sadness Vs. Neutral.

(c) Neutral Vs. Happiness.

(d) Disgust Vs. Happiness.

(e) Disgust Vs. Happiness.

Fig. 2 This figure shows images with common patterns that evoke a contrasting emotional response in humans. The task for the machine is to create a model that correctly elucidates the emotional response

Fig. 3 The idea of happiness is portrayed in OASIS with the above representative set of images. The images have no common set of patterns and we express the associate emotion through metaphors

in the box is an analogy in which everyone has a box that only the owner can see, and no one can see into anyone else's box. Each person describes what he or she sees in the box as a "beetle." The paradox is that whatever represents the beetle cannot have a part in the language game since the thing in the box could be changing all the time, like our emotions, or there might be different things in everyone's box, or perhaps nothing at all in some of the boxes.

4.3 Abductive Reasoning in Brain Programming

This section explains how a new strategy was devised by merging two proposals (Fig. 1d and e) into the flow chart depicted in Fig. 5. Note that we follow a balanced strategy between deductive and inductive reasoning like in previous work about BP. The last work depicted in Fig. 1e successfully provides a methodology capable of separating the foreground from the background on datasets conceived to test salient object detection systems. This knowledge provides a symbolic representation that encapsulates the best possible explanation for what we observe in an image. This

Fig. 4 The idea of sadness is portrayed in OASIS with the above representative set of images. We express the associate characteristics in metaphysical terms since it seems implausible to reach it without valid inferences and using only empirical data

program (rule) derives directly from observations. Abductive reasoning was applied when combining a segmentation step with the saliency map to produce the salient object. The idea is to incorporate the output of such a process (salient object detection) into the AVC model. An intuitive explanation of the abductive reasoning mechanism is as follows: an abductive hypothesis explains a phenomenon by specifying enabling conditions (as a special case) for it. If we want to explain, for example, that the light appears in a bulb when we turn a switch on, an inductive explanation resides in the experience of happening hundreds of times in the past, whereas an abductive explanation bases the analysis on terms of the electric current flowing into the bulb filament [7]. In our example of image classification, the inductive hypothesis appears when we attempt to classify the input through the AVC model since we use the whole image for the computation. However, when we incorporate the salient object detection, we can supply an explanation since we have separated the background from the foreground; thus we can focus more precisely on the part of the image that the system constraints in the computation. This analogy is different from generalization as well as transfer learning. Generalization looks for the ability to achieve good performance under the input of new information, and transfer learning attempts to

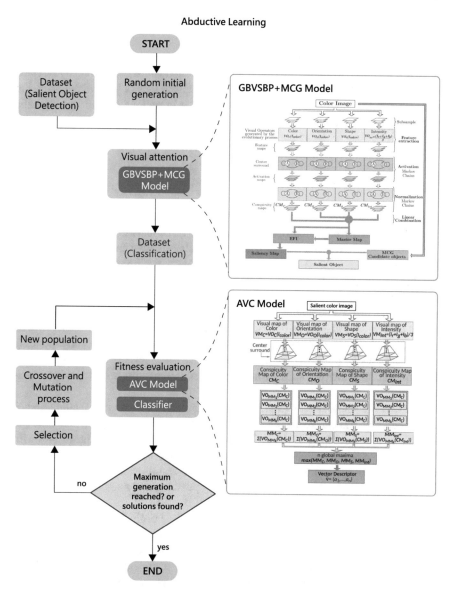

Fig. 5 Flowchart of the abductive brain programming strategy

modify a current model to adapt it to a new problem.[4] Abduction is an inference to the best explanation; epistemologically, we explain the origin of a new hypothesis by abduction. In other words, an inference based on experience toward the specification of a particular goal while taking chances and making the best of ignorance [1]. Here, knowledge encapsulated in a program/rule for a visual task is applied/connected to another process, similar to the idea of modular brain regions to achieve a specific case. This analysis can take us into an era of resilient system design [5].

5   Results

We propose to attempt to solve with GP and other methodologies (inferential knowledge) the visual Turing test proposed in [19]. The test consists of an image database containing pictures collected by psychologists and normative answers made through a carefully designed process to define a ground truth. Preliminary results show the inability to obtain satisfactory results after probing 40 combinations of five different CNNs (convolutional neural networks) and eight optimizers. The study reflects the problem of current ML methodologies. Indeed, the system memorized all images during the learning stage, see Fig. 6. However, in the testing stage, the score reveals a severe problem since predicted valence and arousal values and the corresponding loss across epochs point to a significant difference that reveals CNNs as useless.[5] The correlation principle of DL works with data patterns, not with thoughts-ideas-concepts.

Figure 6 shows the results (training and testing) of three different methodologies using accuracy and F1 score. The column on the left portrays AVC statistics, in the middle corresponding to the AVC + VA, and on the right the SqueezeNet with AdaMax. Note that the neural network mimics human behavior through the information provided in the training set. However, the result drastically drops off with the testing data. The number of images having puzzling and cognitively demanding tasks is limited compared to the total number of images.

On the other hand, BP manages to puzzle out the training set without dropping its performance in testing. Figure 7 shows the superposition with the Fibonacci sequence, and the result is astonishing. The Golden Ratio ($\phi = 1.618\ldots$) is often called the most beautiful number in the Universe. It appears almost everywhere, starting from geometry to the human body itself. Rennaissance artists called this "The Divine Proportion," and the Fibonacci series appears if we divide a term greater than two by a term preceding it. The Fibonacci spiral is a composition guide that creates a perfectly balanced and aesthetically pleasing image in photography. We hypothesize that images in the dataset follow this principle, and when we adopt the program that

---

[4] The literature treated abduction, retroduction, transduction, and even few-shot learning as similar concepts. While tempting to do so, the advice is to study them as different concepts.

[5] CNNs base their computation on correlations, which is why they are invariant to translation and can store a large amount of data.

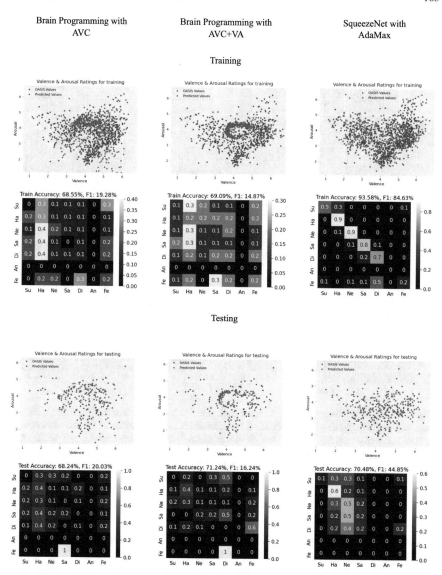

Fig. 6 Machine emotion mapping. Emotion mapping of OASIS images using Brain Programming and a convolutional neural network (Squeezenet with Adamax). Brain Programming's symbolic learning pattern shows us how the machine was capable of abstracting complex concepts and applying them to unseen images. Meanwhile, the statistical learning of the CNN is an example of how a machine can memorize abstract ideas but fail to use them in unknown environments. BP's behavior is reminiscent of the question posed to Turing, can machines think? While the machine's thought process is different from that of humans, this does not mean that the computer is not thinking

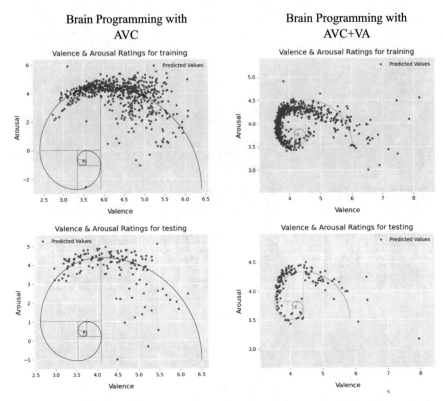

Fig. 7 Machine emotion mapping. A closer look into the pattern formed by Brain Programming when training with the OASIS dataset. If we look closely, the geometry behind the scattered valence and arousal ratings creates a spiral very similar to the one formed by Fibonacci's sequence. This pattern is maintained despite architecture variations and even applied to unseen testing images. The impact of visual attention is also graphically depicted by sharpening the initial shape formed by the AVC

computes the salient object detection, the final result fits this beautiful pattern. This reasoning is a way to understand that the machine/human creates thinking about the OASIS benchmark. The Rule of Thirds and the Golden Ratio materialize through valence and arousal values and can help create/explain a composition/emotion that will draw the eyes/pointer to the essential elements of the photograph. Someone may think that the results did not match the proposed curve; however, we can call to mind some natural phenomena (hurricanes seen from space) that did not perfectly match the Fibonacci sequence but were used to expose the pattern. Nautilus shell is a figure that our outcomes match better for the graceful spiral curve, creating controversy about how accurate it is in following the golden spiral. Understanding that our data is scarce, we do not expect to cover the whole shell, but the vital thing to notice is that independently of what model we are using (AVC or AVC + VA), the output is the same during training and testing.

6 Conclusions

This chapter deals with inferential knowledge and how it is possible to adapt/extend the definitions from social sciences to computer science to construct new knowledge by mimicking thinking in the machine. After an extensive analysis of myths and open issues about GP, ML, CV, and EC, we introduce our approach to synthesizing programs using inferential knowledge and, more specifically, abductive reasoning. BP is a methodology that incorporates domain knowledge at a high level through the idea of templates and at a lower level by selecting the best set of functions and terminals. The resulting symbolic programs provide consistent outcomes for a problem requiring non-contingent representations from the content's viewpoint. The lack of adequacy to a high range of image variation for a single concept is why CNNs fail to model OASIS pictures. In the future, we would like to explore new representations regarding other dimensions, like illumination, that has received little attention from CV and neuroscience communities.

References

1. Andrade, E.: Natural abduction: the bridge between individuals choices and the production of evolutionary innovations. Signs **5**, 112–146 (2011)
2. Aquinas, T.: Summa Theologica I q 79 a 10. Gilby, T. (ed.) OP, vol. 60. Blackfriars, Cambridge (1966)
3. Choi, C.Q.: 7 Revealing Ways AIs Fail. IEEE Spectrum October, pp. 42–47 (2021). https://spectrum.ieee.org/ai-failures
4. Clemente, E., Olague, G., Dozal, L., Mancilla, M.: Object Recognition with an Optimized Ventral Stream Model using Genetic Programming. Applications of Evolutionary Computation, LNCS, vol. 7248, pp. 315–325 (2012) . https://link.springer.com/chapter/10.1007/978-3-642-29178-4_32
5. Cotta, C., Olague, G.: Resilient Bioinsired Algorithms: A Computer System Design Perspective. Applications of Evolutionary Computation, LNCS, vol. 13224, pp. 619–631 (2022). https://doi.org/10.1007/978-3-031-02462-7_39
6. Dozal, L., Olague, G., Clemente, E., Hernández, D.E.: Brain programming for the evolution of an artificial Dorzal stream. Cognit. Comput. **6**, 528–557 (2014). https://doi.org/10.1007/s12559-014-9251-6
7. Flach, P.A., Kakas, A.C. (eds.): Abduction and Induction: Essays on their Relation and Integration. Applied Logic Series. Springer-Science+Business Media, B.V. (2000)
8. Fusing, H., Chen, C., Hsieh, Y.C., Farrell, P.: Lewis Carroll's doublets net of English words: network heterogeneity in a complex system. PLoS One **9**(12), e114177 (2014). https://doi.org/10.1371/journal.pone.0114177
9. Hernández, D.E., Clemente, E., Olague, G., Briseño, J.L.: Evolutionary multi-objective visual cortex for object classification in natural images. J. Comput. Sci. **17**(Part 1), 216–233 (2016). https://doi.org/10.1016/j.jocs.2015.10.011
10. Hofman, M.A.: Chapter 18 - design principles of the human brain: an evolutionary perspective. Progr. Brain Res. **195**, 373–390. https://doi.org/10.1016/B978-0-444-53860-4.00018-0
11. Ibarra-Vazquez, G., Olague, G., Chan-Ley, M., Puente, C., Souberville-Montalvo, C.: Brain programming is immune to adversarial attacks: towards accurate and robust image classification using symbolic learning. Swarm Evolut. Comput. **71**, 101059, 17 pages (2022). https://doi.org/10.1016/j.swevo.2022.101059

12. Koza, J.R.: Genetic Programming: On the Programming of Computer by Means of Natural Selection. MIT Press, Cambridge (1992)
13. Koza, J.R.: Human-competitive results produced by genetic programming. Genet. Program. Evolv. Mach. **11**, 251–284 (2010). https://doi.org/10.1007/s10710-010-9112-3
14. Luger, G.F.: Artificial Intelligence: Structures and Strategies for Complex Problem Solving. Addison-Wesley (2002)
15. Maynard-Smith, J.: Natural selection and the concept of a protein space. Nature **225**, 563–564 (1970). https://www.nature.com/articles/225563a0.pdf
16. Mitchell, T.M.: Machine Learning. McGraw-Hill (1997)
17. Olague, G.: Evolutionary Computer Vision: The First Footprints. Springer (2016)
18. Olague, G., Clemente, E., Hernández, D.E., Barrera, A., Chan-Ley, M., Bakshi, S.: Artificial visual cortex and random search for object recognition. IEEE Access **7**, 54054–54072 (2019). https://doi.org/10.1109/ACCESS.2019.2912792
19. Olague, G., Olague, M., Jacobo-Lopez, A.R., Ibarra-Vázquez, G.: Less is more: pursuing the visual turing test with the Kuleshov effect. In: IEEE/CVF Conference on Computer Vision and Pattern Recognition Workshops (CVPRW), pp. 1553–1561 (2021). https://doi.org/10.1109/CVPRW53098.2021.00171
20. Olague, G., Menendez-Clavijo, J.A., Olague, M., Ocampo, A., Ibarra-Vázquez, G., Ochoa, R., Pineda, R.: Automated design of salient object detection algorithms with brain programming 35 pages (2022). arXiv:2204.03722v1. https://doi.org/10.48550/arXiv.2204.03722
21. O'Neill, M., Vanneschi, L., Gustafson, S., Banzhaf, W.: Open issues in genetic programming. Genet. Program. Evolv. Mach. **11**, 339–363 (2010). https://doi.org/10.1007/s10710-010-9113-2
22. O'Neill, M., Spector, L.: Automatic programming: the open issue? Genet. Program. Evolv. Mach. **21**, 251–262 (2020). https://doi.org/10.1007/s10710-019-09364-2
23. Rich, C., Waters, R.C.: Automatic programming: myths and prospects. Computer **21**(8), 40–51 (1988). https://doi.org/10.1109/2.75
24. Rosenfeld, A.: Religion and the robot. Tradit: J. Orthodox Jewish Thought **8**(3), 15–26 (1966). http://www.jstor.org/stable/23256081
25. Rusell, S., Norvig, P.: Artificial Intelligence a Modern Approach, 3rd edn. Prentice Hall (2010)
26. Salisbury, F.B.: Natural selection and the complexity of the gene. Nature **224**, 342–343 (1969). https://www.nature.com/articles/224342a0.pdf
27. Samuel, A.L.: Some studies in machine learning using the game of checkers. IBM J. Res. Develop. **3**(3), 210–229 (1959). https://doi.org/10.1147/rd.33.0210
28. Turing, A.M.: Computing machinery and intelligence. Mind **LIX**(236), 433-460 (1950). https://doi.org/10.1093/mind/LIX.236.433
29. Wittgenstein, L.: Tractatus Logico-Philosophicus, first English-language edition, Harcourt, Brace & Company, Inc (1922)

Life as a Cyber-Bio-Physical System

Susan Stepney

Abstract The study of living systems—including those existing in nature, life as it could be, and even virtual life—needs consideration of not just traditional biology, but also computation and physics. These three areas need to be brought together to study living systems as cyber-bio-physical systems, as *zoetic systems*. Here I review some of the current work on assembling these areas, and how this could lead to a new Zoetic Science. I then discuss some of the significant scientific advances still needed to achieve this goal. I suggest how we might kick-start this new discipline of Zoetic Science through a program of Zoetic Engineering: designing and building living artefacts. The goal is for a new science, a new engineering discipline, and new technologies, of zoetic systems: self-producing far-from-equilibrium systems embodied in smart functional metamaterials with non-trivial meta-dynamics.

1 Introduction

It is certainly the case that "Nothing in biology makes sense except in the light of evolution" [1]. Although this evolutionary light is necessary, it is by no means sufficient for such sense making. In order to make sense of life,[1] and of living systems, we also need the light of physics and computing.

Here I survey the advances made in bringing these lights of computing and physics to bear on living systems and suggest a new Zoetic Science that fully integrates these

---

[1] I will not get into definitional questions of what makes systems 'really' 'alive' [2]. However, consider "living organisms are those material systems that are able to manipulate information so as to produce unexpected solutions that enable them to survive in an unpredictable future", and "life as a process that enables material systems to manipulate, create, and accumulate information" [3].

S. Stepney (✉)
Department of Computer Science, University of York, York, UK
e-mail: susan.stepney@york.ac.uk

© The Author(s), under exclusive license to Springer Nature Singapore Pte Ltd. 2023
L. Trujillo et al. (eds.), *Genetic Programming Theory and Practice XIX*,
Genetic and Evolutionary Computation, https://doi.org/10.1007/978-981-19-8460-0_8

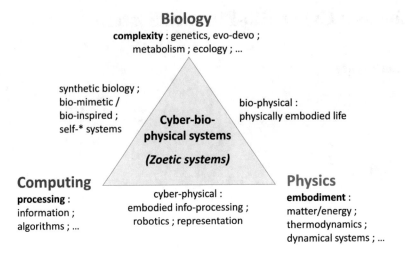

Fig. 1 The triangle of life, combining biology, computing, and physics in cyber-bio-physical systems

components. I suggest the way forward to developing this science is through Zoetic Engineering case studies, with an example of starting from a basis of swarm robotics and morphogenetic engineering.

2 The Triangle of Life

I argue that the science and engineering of life, of living systems, requires more than the study of biology alone. Biology as it stands today studies *life as we know it* on Earth; exobiology studies (currently hypothetical) life in the rest of the Universe. Artificial Life (ALife) studies *life as it could be* more generally [4], including theoretical, computational, and experimental approaches, and with consideration of fields and aspects beyond that of core biology itself, particularly (i) the physical embodiment of the living systems, and (ii) the informational and computational aspects of living processes. Even the study of life as we know it needs to include these physical and computational aspects for a fully developed approach. I encapsulate these aspects conceptually in Fig. 1.

The vertices in Fig. 1 represent the individual fields that together are needed to capture the complexity of living systems. In Sects. 2.1–2.3, I pick out a few aspects of each discipline that are most relevant to the topic of life and life-like systems. The edges in Fig. 1 represent pairwise cross-disciplinary interactions between the individual fields. In Sects. 2.4–2.6, I pick out a few relevant aspects of these interactions. The face in Fig. 1 represents the combination of all three aspects brought together as cyber-bio-physical systems; see Sect. 2.7.

2.1 B: Biology

Biology is a science of complexity, the complexity of the living world. Since it has only one instance for study (life on earth), it typically does not address questions of what is necessary for life, and what is merely contingent of its realisation on this planet, and given its specific evolutionary history. One of the goals of the field of ALife is to broaden this scope.

The topics of biology can be broken down into four approximate levels (below). These levels are often studied in isolation, although cross-level effects, such as evo-devo (the interaction of evolutionary and developmental processes) and evolutionary ecology, are considered. One key constraint of natural living systems is that individuals and all their evolutionary ancestors are viable throughout their processes of birth, maturation, and reproduction: there is a continuous chain of life.

The entire subject is incredibly complex, and there are many exceptions to any rule; specific models tend to be preferred over abstract approximate ones [5].

Substrate. The lowest level considered is the substrate of biomolecular materials and processes. This includes biochemical processes of manipulation and construction of the substrate: metabolism (catabolism, the breaking down of large molecules and releasing of energy, and anabolism, the building up of more complex molecules, using energy). It also includes reproduction and replication of the substrate.[2] This is based in the field of molecular genetics, of the 'control program' of transcription, translation, and replication of DNA.

Individual. The individual level is that of a single whole organism. This includes sensing, (re)acting, moving, adaptation, and learning, including neural cognition (in animals) and adaptive immunity (in vertebrates). For multicellular organisms, it includes the process of ontogeny/development/growth from single cells to mature organisms to death, again mediated by the genetic program. There is no single solution; organisms adapt to and are shaped by their context or environment:

> Other minds, other worlds from the same monotonous and inexpressive chaos! My world is but one in a million alike embedded, alike real to those who may abstract them. How different must be the worlds in the consciousness of ant, cuttlefish, or crab! [6, Chap. IX]

Population. This is the species level. It includes evolution: how the genetics changes. It covers populations of individuals both in terms of belonging to a species, and also as 'superorganisms' (for example, social insects) acting as a higher level individual. It includes ideas of major transitions in complexity [7, 8], open-endedness [9, 10], and goes meta with evo-evo (the evolution of evolutionary processes) [11].

System. The highest level considered is of an ecosystem of multiple species, and their interactions with each other (such as competition for shared resources, predation and

---

[2] A distinction is drawn between reproduction (of the physical machinery, the cell) and replication (of the informational content, the DNA) [3]; many abstract models (for example, basic evolutionary algorithms) however do not separate these processes.

food webs, cooperation, and symbiosis) and with the physical environment (exploiting and creating niches). Ecosystems exist on multiple scales, from communities to the entire biosphere.

2.2 P: Physics

Physics is a science of (relative) simplicity, that of matter and energy behaving in space and time, of embodiment.

2.2.1 Matter

Matter forms the substrate material of living organisms and of their physical environment. The biological processes are embodied in the physical material and are subject to its constraints (the processes it limits) and its opportunities (the processes it supports, enables, or performs). For life on earth, many of these processes are chemical in nature and non-ergodic.

This substrate material occurs in highly evolved complex structures and molecules that are not observed in non-living matter. The existence of suitably complex molecules can be used as a biosignature, either directly of the evolved life itself or indirectly of an artefact engineered by evolved life [12]. (Free oxygen is also considered to be a sign of life.) As well as providing the substrate of biological processes (metabolism, growth, reproduction, etc.), the matter is used to store the energy (typically as chemical energy) needed to drive those processes.

2.2.2 Energy and Entropy

Thermodynamics is the physics of energy conservation and entropy production. Energy is needed to drive biological processes; the flux of sunlight can be accessed directly by plants; this flux needs to be converted to chemical energy for use by other organisms. Chemical energy can be stored, which allows for energy use when the flux is not available (for example, in shadows and caves, or at night), or not instantaneously sufficient for certain processes.

Energy flux is lost to the system if not used or stored; matter is constantly recycled in ecosystems [13]. Energy is conserved overall, but it gets degraded from free energy to forms that cannot be used to do work: entropy increases. Life, despite being able to locally decrease entropy by creating order and structure, nevertheless overall increases the rate of entropy production [13]. As Russell [14] puts it:

> the *raison d'être* of life is to dissipate energy and produce heat and waste, the exhausts from biosynthesis: life is an entropy generator. The more entropy an organism can generate, the more successful it is: for a while.

2.2.3 Time and Dynamics

Physical systems are often modelled as dynamical systems: a set of differential equations (DEs) modelling the time evolution of the state of the system as a trajectory through state space.

Non-linear differential equations are typically not analytically solvable. Hence, in earlier decades, out of necessity it was traditional to study linear dynamical systems, typically through the wave equation, diffusion equation, Poisson equation, Fourier transforms, and the like. Linearity has the great advantage of being additive, allowing solution techniques such as Green's functions. Weak non-linearity is handled through power series expansions in terms of small deviations from linearity.

Aside from quantum mechanics, however, the bulk of interesting systems in physics and beyond (from general relativity to chemical reaction kinetics and epidemics) are strongly non-linear:

> using a term like nonlinear science is, as the noted pioneer in experimental mathematics Stanislaw M. Ulam has observed, like referring to the bulk of zoology as the study of non-elephant animals [15].

This field of non-linear science has been developing, due to advances in the analysis of non-linear dynamical systems. In addition to advances in numerical solution, the field of non-linear dynamical systems encompasses the concepts of bifurcations (sudden changes in behaviour from a small change in the parameter value), chaos and sensitive dependence on initial conditions, classes of attractors (regions of state space where dissipative system trajectories end up) including strange attractors (where nearby trajectories eventually diverge, yet remain in the attractor), fractals, and more. See, for example, [16, 17].

Many of these systems are multi-dimensional, involving coupled differential equations, for example, Lotka-Volterra systems and chemical reaction networks. A multi-dimensional system can be represented as a network, where the nodes represent state variables, and the edges link to variables that affect the time evolution of the node. This leads to the concept of dynamics on networks.

In a spatial system, each point of space can be considered a state variable, affected by the neighbouring points. This leads to partial differential equations (PDEs), in contrast to ordinary differential equations (ODEs) that have derivatives with respect to a single variable, usually time. PDEs describe infinite dimensional systems with a spatial topology linking the state variables. Since space is continuous, PDEs include derivatives with respect to this continuous space, as well as with respect to time. The classic linear examples are Poisson's equation and the diffusion equation. A non-linear example is reaction-diffusion systems, such as those producing Turing patterns [18].

2.3 C: Computing

Computing is the science of information and processing. The subject includes a range of hierarchical structuring and software architectures, including layering software using virtual machines (VMs). One of the key abilities computational systems have that is crucial for life is the ability of *reflection* [19]: code can potentially refer to itself, run itself, and even modify itself.

Computing is often presented as a sub-branch of mathematics, a purely abstract discipline. However, all information, and all processing, have to be embodied somehow in the physical world, which places constraints on the possibilities. Computing can be viewed as a natural science [20].

The classical computation model, of the Turing Machine (or one of its equivalent formulations) is a symbolic, discrete time, discrete space, deterministic, 'ballistic' (dependent only on its initial condition; closed to inputs during execution), halting, sequential model. Other computational models exist, which break one or more of these assumptions, including stream processing and interactive computing [21] (which are open models that include communication with the environment during processing), analogue computing (where a physical model of the system under study is used as a computational analogue), general purpose analogue computing [22] (where mechanical or electrical circuits are used as physical analogues of an ODE model of the system), cellular automata (a form of parallel processing), Artificial Neural Networks (inspired by neural processing in the brain), and more.

These other computational models can be *simulated* by a classical digital computer—they are not uncomputable models [23]—but the point is about naturalness and explanatory power: the model should fit what it is modelling. Different models are suited to different requirements of information and processing, such as encoding and decoding information; control systems for sensing and interacting with an environment; cognitive approaches for memory and learning.

Since different models are appropriate for different problems, a complex problem such as life needs to include the composition of the relevant models, or some unification of them [24]. In particular, hybrid systems include a combination of both discrete and continuous models.

2.4 B-P: Bio-physical Systems

'Bio-physics' can simply mean using physics-based approaches to studying biological systems. While some such approaches may work, living systems are qualitatively different from the kinds of systems typically studied in physics, and so it is important to avoid 'physics envy' or 'physics chauvinism' in this cross-disciplinary topic.

Here I consider the topic to be the study of how the laws of physics constrain and enable life-like processes. It is the study of how physical embodiment affects life, recognising that not all features of life are purely due to genetics, but that some come from the physical properties of the material substrate itself [25].

This embodiment is most recognised at the substrate level, the level of molecular biology, biochemistry, chemical reactions and energetics, and the like [3]. Stiffness and supercoiling of DNA, protein folding, molecular diffusion in the cell, fluid dynamics of moving cells, and more, all these are physical properties. These properties result from highly evolved complex substrate molecules; such molecules can have very different properties from the non-evolved materials typically studied in physics. For example, entropy is usually conceptualised as a disorder. However, entropy increase can have non-intuitive outcomes for complex molecular systems, for example, of promoting segregation in certain cases:

> under strong confinement conditions, topologically distinct domains of a polymer complex effectively repel each other to maximize their conformational entropy, suggesting that duplicated circular chromosomes could partition spontaneously [26].

There are physical resources constraints at all levels, particularly those imposed by the environment, including sources of energy, materials, and niches.

2.5 C-B: Cyber-Bio Systems

'Computational biology' is the use of computers to analyse biological data and model biological systems and is not the topic of interest here. Rather, the topic is 'biological computation', the use of computational concepts as ways to explain and model biological processes, with the idea that some or many of these processes are intrinsically computational in nature.

When viewing living matter computationally, it can be approached in classical computing terms, from logic gates [27] to operating systems and beyond [3] (although there are differences between computer and cellular architectures [28]), or through using a broader concept of computing with models more suited to bio-substrates [29]. Computational analogues of biological processes may be suggestive; for example, partial evaluation [30] may be seen as a computational analogue of the Baldwin effect [31].

2.5.1 Information Processing

The clearest example of information and processing in biology is that of DNA and the genetic machinery that decodes and processes that information. This can be treated in an information-theoretic manner, including the role of error correction. The complexity of the information can be measured in terms of the mutual information between the genome and its environment; evolution increases this mutual information [32].

Different domains of life (bacteria, archea, and eukaryotes) have different 'operating systems', or cellular machinery, and so a gene or genome transplanted from one to another domain (or even within different clades of one domain) will not necessarily 'execute' in the same way [3]. The cellular machine reads the program (DNA), but can also under certain circumstances write/change it, allowing the program to change (as in self-modifying code).

2.5.2 Developmental Processes

Danchin [3, 33] views biological development as an algorithmic process, as iteration in space (parallelism) and time, constructing machinery that can replicate itself. He views this ability as one of the defining features of life: "the general principles for the construction of a self-replicating machine have nearly always overlooked the need for compartmentalization and metabolism" [33, p. 253].

Development as a form of algorithmic 'unfolding' includes complexity in the form of Bennett's logical depth [34]. Deep algorithms take time to unfold, and the processes of development (or evolution) can require that time.

Lindenmayer systems (L-systems) were developed initially to model growth processes in filaments [35, 36] and have been extended to model a host of plant-like developmental processes [37]. They are a form of generative grammar, where symbols in a string modelling the growing organism are rewritten in parallel to each generation, representing the parallel growth processes of each part of the organism. The field of Morphogenetic Engineering [38] explores the interplay of bottom-up self-organisation and top-down architecture to form programmable 'self-architecturing systems', typically in simulation, although the vision is of physical embodiment.

2.5.3 Synthetic Biology

Synthetic biology is the engineering side of genetic biology [39]. Viewing the genetic machinery as performing computations in a cell, with proteins as part of the signalling mechanism, it is natural to seek to engineer-in specific computations, for example, to manufacture particular proteins.

The computational model typically used in this case is based on genetic circuits modelled as discrete boolean circuits. However, DNA expression and control include processes outside this relatively simple model, such as methylation (extra-state information) and supercoiling (further regulation mechanisms) [40]. Additionally, real genetic machinery is 'leaky' (non-boolean) and stochastic. More sophisticated computational models are needed to better capture the relevant biological processes involved.

2.5.4 Bio-inspired Algorithms

Biological processes can themselves be taken as inspiration for a range of computational algorithms, such as metaheuristic optimisation. Evolutionary algorithms [41,

42], Particle Swarm Optimisation (PSO) [43, 44], Ant Colony Optimisation (ACO) [45], and a range of Artificial Immune algorithms [46] are used for search and optimisation; a range of artificial neural network [47, 48] algorithms and other biological network-inspired algorithms [49] are used for classification, prediction, and control.

Population-based optimisation algorithms can be classed as a form of *swarm intelligence* [50] where the desired outcome is a property of the 'fittest' individual in the population. In 'true' swarm intelligence algorithms, however, the result is due to the collective behaviour or configuration of the entire population (inspired by such biological behaviours as termites building a structure, or ants forming a bridge with their bodies); the swarm can be thought of as a single 'superorganism'.

Despite the range of inspirations used to develop such algorithms (some argue, much too wide a range [51]), there are few abstract underlying models. It might be argued that all the population-based optimisation algorithms are at heart the same [52]: they have a population of individuals, a fitness measure, a way of breeding new individuals based on this fitness, and a balance between exploration to find new solutions and exploitation of good solutions; they differ only in the details. Novelty search [53, 54], by eschewing exploitation for pure exploration, is arguably the most innovative advance in population-based algorithms in recent times. Similarly, the various network-based algorithms (neural, metabolic, signalling, and genetic regulatory) have a single conceptual model [55] and can be considered as population-based (nodes) with the addition of a relational structure (edges).

Such algorithms, although inspired by biology, are not particularly biologically plausible: they typically have much smaller populations, smaller 'genomes' (individual complexity), less sophisticated mechanisms for breeding new individuals, and less rich concepts of fitness than does biology itself. The presence of such a small set of abstract models, realised in such a broad range of bio-materials and mechanisms, demonstrates that much richness can be achieved using essentially the same mechanisms in a wide variety of contexts and scales.

2.6   C-P: Cyber-Physical Systems

'Cyber-physical' systems are typically embedded systems, where a conventional computer is embedded in and controls a physical system such as a robot or an 'intelligent' building. Here, however, I use the term in the context of a more life-like system, where the (unconventional) computational and information-processing aspects are embodied in a physical system: the processing 'brain' and the physical 'body' are intimately entwined, sensing, and interacting with an external environment in a feedback loop, as in cybernetic systems.

2.6.1 Embodiment

All computing, even classical computing, is embodied. Information is physical [56], embodied in material structure, from magnetic domains on a hard disk to sequences of bases in DNA. Processing is embodied in the physical dynamics of structured material, from flows of electrons in engineered circuits to oscillations of substrate properties in *in materio* reservoir computing [57, 58] and to feedback control of physical configurations in soft robotic bodies.

Classically, algorithms or programs are used to define and control the precise processing that occurs. Embodied processing may have no explicit program: it may be encoded in the values of certain material properties and configurations of the physical system. These may be engineered through some explicit learning process, as in reservoir computing, or evolved, as in natural organisms.

The focus of embodiment is often on embodied intelligence: how the brain and body work together [59] and influence each other [60]. However, it can cover close coupling between 'processing' and 'physical system' at all levels [61], not only high-level cognition. Although it is generally recognised that body morphology provides an important contribution to cognition and control, there is some dispute about whether there is any specifically computational contribution [62].

2.6.2 Information, Physics, and Life

Some authors identify information as a fundamental concept in physics [63]. Rather than subsume yet another concept under the domain of physics, however, here I consider computing to be a separate domain. The form information processing and computing take in an embodied context is typically not the same paradigm as in a conventional symbolic digital computer, but tends to involve more unconventional models and implementations of computing. In such cases, it is necessary to decide when a physical system is simply 'doing its thing', merely obeying the laws of physics, and when it is in addition also computing.

In order to determine when a particular physical process is performing computation, Horsman et al. [20, 64] use the definition: "the use of a physical system [the computer] to predict the outcome of an abstract evolution [the computation]." This use requires a representation relation between physical and abstract, which is used by a representational entity [65, 66]. One feature of this definition is that the representation relation is essentially arbitrary and can be realised in different ways independent of the physical substrate; see also Sect. 3.7.

The representational entity may be performing *intrinsic* computing internally (for example, by a bacterium to navigate towards food) or *extrinsic* computing by exploiting some external device (for example, a person using a PC). Thus, in addition to the argument that life requires computing (information and processing, Sect. 2.7), it can be argued that computing requires life, in order to *represent* information and processing, either embodied in itself, or by employing an artefact constructed to embody them.

2.6.3 Constraints and Limits

Computing is not merely a subset of mathematics, but it is not merely a subset of physics either [20]. However, given that computing is embodied in, and performed by, a physical system, it is clearly constrained by the laws of physics. This is just as true of classical computing, but tends to be neglected by theoretical computer science.

For example, there are physical limits on the computational speed of quantum gates, determined by fundamental physical dynamics [67]. Lloyd [68] has estimated the physical limits on the processing power of a one litre, one kilogram computer in terms of the values of fundamental physical constants. The Landauer principle [69] relates physics and information by linking to thermodynamics and irreversibility; Bennett [70] provides an excellent summary.

Even if a living system is not operating at the extremes of such fundamental constraints, there are still limits to consider because of its physical embodiment, including the source and amount of energy available to drive the dynamics; the source and quality of material to build and maintain the system; the ability to dispose of entropy in the form of waste heat and material.

2.7 Pulling It All Together: C-B-P

Living and life-like systems require consideration of (at least) the whole triangle of disciplines (Fig. 1):

- computing: information, processing, and 'intelligence';
- physics: material embodiment (of information and processing) in the structured matter, providing constraints and opportunities;
- biology: adaptation (evolution, learning), self-construction (growth, assembly), self-maintenance, open-endedness.

However, the computing, physics, and biology involved are not in their classical form. When we put together vertices to make the edges of the triangle, considering the overlap of pairs of disciplines, the science of each vertex is expanded and changed in the process. And these edges are interpreted differently in the consideration of life itself from the usual uses of the terms. Bringing all three together for a full science of the living requires further modifications and extensions.

2.7.1 Arbitrary Symbolic Relationships

Danchin [33] argues biology comprises *symbolic* relationships (for example, the genetic code) that, although constrained by physics, are nevertheless arbitrary (the genetic code could be different), so are not deducible from or reducible to physics. "Replaying the tape" of evolution [71] could well have resulted in a different genetic code or differences in other symbolic relationships in biology. "The objects that make biological functions happen often have no mechanical relationship with them; they are only their mediator, their *symbol*." [33] This symbolic nature has consequences for the underlying physics: it permits, constrains, and determines certain *classes* of symbols, but does not constrain the actual instances chosen. Even this constrained space is vast, and the realised actuality is just a small, arbitrary subset of this, chosen by evolution. We know that evolution is adept at finding and exploiting small differences and 'bugs' [72], so it might be that there is in fact some optimal symbol choice, but the richness of our biological world suggests there is a broad exploitable plateau around any such optimum or multiple optima.

Danchin [33] points out that this symbolic property allows "symbolic mediation" between quite different domains, such as representation within DNA and its realisation within proteins, and so the detailed nature of the underlying physical processes can be separated from the abstract symbolic processes. This separation is a form of emergence. Laughlin [73] points out that when emergent properties are insensitive to the substrate (as in this case), it is not possible to draw conclusions about the substrate from them. So we should not expect to be able to draw conclusions about the physical substrate from observing the symbolic biological processes.

2.7.2   Example Domains

Here I briefly discuss a few example domains that include the whole cyber-bio-physical spectrum to some degree: embodied cognition, swarm robotics, and morphogenetic architecture.

Embodied cognition. This area combines computationalism [74]—the brain as information processor—and cyber-physical embodiment—computational processes embodied in physical devices, or 'bodies'—interacting with an environment; the interaction provides a mechanism for symbol grounding. The focus is on cognition; the living body is a given. Cognition need not imply neural-style processing: Cohen [75] argues that the adaptive immune system is a cognitive system.

Swarm robotics. Swarm robotics combines the bio-inspired algorithms of swarm intelligence with embedded cyber-physical robotic systems [76]. The concept of using a multitude of small, simple, and (relatively) cheap robots, rather than engineering a single large, complicated, expensive device to do the same tasks, has led to the idea of "fast, cheap, and out of [top-down] control" systems [77, 78].

Swarm robotics research includes a range of biological concepts, such as morphogenesis (robot swarms self-organising into emergent shapes) [79], evolution [80], and open-endedness [81].

Current research tends to focus on the bio-inspired algorithms, and traditional embedded computational robot controllers. A more embodied computational control can be seen through research into soft robots and nanorobots.

Morphogenetic Architecture. The field of morphogenetic architecture [82] uses bio-inspired ideas of morphogenesis for designing buildings. A form of swarm robotics can be used to build structures, either directly, as 'smart bricks' that self-assemble, or indirectly, as 'smart termites' that assemble less smart materials [83]. While swarm robotics is based on biological analogues of animal populations, morphogenetic architecture is also based on plant behaviours [84]. Other aspects of 'soft living architecture' [85] include the homeostatic functioning of the completed building. Plants can also form the inspiration for novel sensors and actuators [86, 87] for use in 'smart' buildings.

2.7.3   Zoetics

Life is a cyber-bio-physical system, but that term does not trip readily off the tongue. The term 'biology' is already taken for the study of naturally occurring living systems, but, as I have discussed above, tends not to cover the entirety of information processing, embodied, life-like systems. The discipline of Artificial Life [4] includes these aspects, but might be thought to exclude natural life. The word 'lyfe' has been coined to describe any system that exhibits dissipation, autocatalysis, homeostasis, and learning [88], in the context of astrobiology; however, we want to include artificial systems.

The adjective 'zoetic' means[3] 'of or relating to life: living, vital' (from the Greek ζωή, zōē, life[4]), and so I use that to refer to all living systems, natural or artificial. In the next section, I overview some topics that are needed for Zoetic Science, then in Sect. 4 I discuss Zoetic Engineering.

3   Zoetic Science

Science, in a nutshell, is studying the world *as it is*, in order to build models of increasing explanatory and predictive power. Here I discuss some of the tools and topics (Fig. 2) that need extensions to deal with the particular properties of living systems; in Sect. 5, I briefly note some other topics not covered here.

3.1   Philosophy

The philosophical underpinnings of Zoetic Science need to be made clear: both how we perform scientific enquiry, and the basis of the underlying subject of study.

---

[3] https://www.merriam-webster.com/dictionary/zoetic.

[4] Not to be confused with the separate, though related, etymology of 'zoology' via Latin from the Greek ζῷον, zōion, animal.

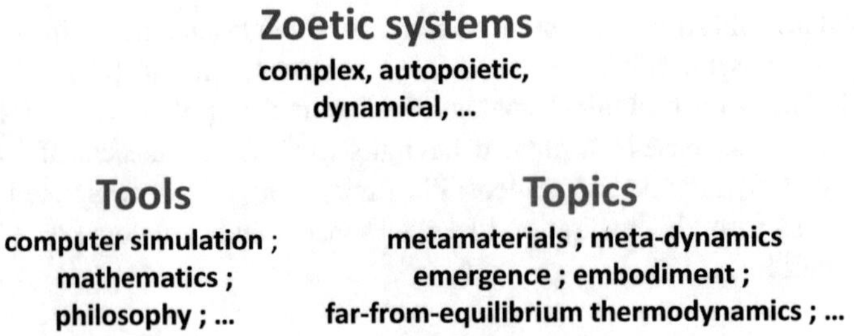

Zoetic systems
complex, autopoietic, dynamical, ...

Tools	Topics
computer simulation ;	metamaterials ; meta-dynamics
mathematics ;	emergence ; embodiment ;
philosophy ; ...	far-from-equilibrium thermodynamics ; ...

Fig. 2 Some of the tools and topics needed within Zoetic science, discussed in Sect. 3

3.1.1   Epistemology

Not all science can be predictive the way physics aspires to be; the life sciences are simply too complex, too messy, and do not have the simplifying assumptions available to many branches of physics, such as well-separated length- and time-scales, identical particles, isolatable systems, pre-defined state spaces, and continuous dynamics. This leads to researchers in these disciplines having different relationships with abstractions, theories, models, experiments, data, and explanations [5, 89].

We need to ensure that Zoetic Science uses an appropriate method and philosophy for its mode of study, and not simply lift existing (often ill-defined, typically inappropriate) approaches from its component disciplines. This will require systematic reflection on the way the science is conducted: a second-order science [90].

3.1.2   Relational View

Leibniz proposed a relational model of space and time. Rather than Newton's absolute model where space and time exist independently, in Relationalism, "spatial and temporal relationships between objects and events are immediate and not reducible to space-time point relations, and all movement is the relational movement of bodies" [91].

A relational view can be applied to complex systems in general; modelled as a graph, the organisational structure (edges) takes priority over the material (nodes) [92]. Hence, we get the tale of the Ship of Theseus, asking if all the components are replaced, is it the 'same' ship? Yes, in a relational view, as the crucial structure has been preserved.

Relations can be first-class objects, allowing relations between relations [93]. Such a view helps explain why emergent properties are insensitive to the underlying material [73]: the emergence is building on the relational structure, not the specific matter that supports it. It also helps with the inclusion of information in a model: information is embodied in structure.

3.1.3   Process View

As Charlotte Perkins Gilman puts it: "Life is a verb, not a noun. Life is living, living is doing, life is that which is done by the organism." [94, Chap. X].

Ingold [95], arguing for an organism-centric biology, says:

> It must be a biology that asserts the primacy of processes over events, of relationships over entities, and of development over structure.

He also quotes Cassirer [96, p. 72], who says[5]

> Organic life exists only so far as it evolves in time. It is not a thing but a process—a never-resting continuous stream of events. In this stream nothing ever recurs in the same identical shape.

Life as a *process* is an essentially temporal, dynamical concept. Living systems embody many processes: they have a lifecycle of becoming, being, ceasing; they have sub-processes during this of maintaining, repairing, growing, adapting, learning, interacting, and more.

This suggests that a process view [97], rather than a substance view, is a more appropriate view of life [98–100]. However, life may be a verb, but it is not a *disembodied* verb. A pure process view may be too extreme: both process and matter are key [101]:

> neither of matter/object nor process/event is ontologically prior to the other; but rather, each is dependent on the other. [...] *(a) matter and objects by nature presuppose the participation in processes or events, and (b) processes and events by nature presuppose the existence of matter or objects.*

3.2   Systems

Rather than focussing on nouns (components) or verbs (processes), we can take a systems view, where the components and processes are packaged together in an integrated and structured (relational) whole.

3.2.1   Systems View

A systems view considers certain processes and material as a coherent whole [102]. A systems view is antithetical to a reductionist view in the following sense. A reductionist view starts by breaking the object of study into its components, in order to understand the components in isolation, then (hopefully) reassemble them to understand the system as a whole. The systems view, on the other hand, starts by examining the context of the system: what is its environment, what does it interact with, what is

---

[5] I give a slightly longer quotation here than appears in Ingold [95].

its history, and how does it behave in and because of its context? Since a key feature of living systems is their adaptability, examining their relationship with the context to which they adapt seems a reasonable point of view.

3.2.2 General Systems Theory

This idea of open systems, of wholes interacting with their environments, was the motivation for von Bertalanffy's development of General Systems Theory [103]. It grew out of thinking of biological organisms holistically. Such an approach is needed for non-linear systems where the parts are closely coupled (the very properties that make something a 'system', rather than an aggregation, a 'mere heap'). It is not just the components, but their relationships, their interactions, and dynamics, that make up a system and give rise to emergent system-level properties not seen in the individual components.

These early systems approaches tend to be produced from a physical rather than computational worldview. As such, they tend to focus more on relationships and patterns of structure, rather than the information and processes driving the dynamics. More modern approaches can take a more computational view.

3.2.3 Systems Biology

Systems Biology takes a system (that is, non-reductionist) view of biology, but the systems considered are mostly confined to the molecular/cellular level. Systems thinking [102] views systems such as these molecules and cells as themselves components and processes of larger systems, leading to a hierarchy of *systems of systems*. This hierarchy is not a pure tree [104]: peer-level systems are also coupled, although more weakly between systems than within systems. DeLanda [105] discusses structures comprising hierarchies and 'meshworks', and how these concepts can be applied across a range of scales and domains.

3.2.4 Autopoietic Systems

Autopoiesis [106] focuses on the organisation (the network of component-producing processes [107]) that makes a living system a 'unity'. For autopoietic ('self-producing') systems, their operation (the processes they perform) produces themselves (the components that embody those processes). Contrast this to an allopoietic ('other-producing') system, whose operation produces something other than what it is made of, and is produced by a system other than itself (for example, a factory machine, producing unrelated widgets, itself built elsewhere). Allopoetic systems are necessarily open since they need input to produce them, they are built; autopoeitic systems are to some degree closed since they build themselves (up to constraints of the second law of thermodynamics). Autopoetic organisation can be substrate-independent: "the same organization may be realized in different systems with different kinds of components as long as these components have the properties which realize the required relations" [106]. There is also work on realising autopoietic organisation in virtual systems [108].

3.2.5   Unnarratable Non-systems

Not all of life's processes are readily considered as systems. Consider the process of evolution. We have given it a name, 'thingified' it, but we do not view it as a system. It is a process within a living system of populations and ecosystems of organisms. Abbott [109] says that we understand the world through explanatory narratives of entities with agency. Parts of the world that do not have suitable structure are unnarratable and hence are not easily understood. He explores this example of evolution as one such process. It may be that complex systems are fundamentally unnarratable. "There isn't a story. It's more like tending a garden, only you're growing it with 10,000 other gardeners."[6] Gardening may be a good metaphor, or model, for thinking about interacting with complex multi-scale living systems [93, 111].

3.3   Mathematics

These complex, self-referential, process-oriented, autopoetic zoetic systems will need advanced mathematical underpinnings to define and model them.

3.3.1   Mathematical Self-reference

Autopoeisis (Sect. 3.2.4) is a circular process: A makes B makes C makes A. The mathematics of life needs to support circularity and self-reference. Such self-referential definitions (which include Russell's paradox [112]) are explicitly excluded in traditional mathematical set theory, through the axiom of foundation (essentially: every definition has to 'bottom out' eventually). It might be true that "the axiom of foundation has played almost no role in mathematics outside of set theory itself" [113], but the traditional set theory has an enormous impact on the way scientists model the world.

   Self-referential definitions are perfectly allowable in non-wellfounded set (or hyperset) theory, which instead includes the axiom of anti-foundation, one form of which was developed to provide the semantics for Process Algebra formalisms in computer science [114]. This alternative form of set theory allows both endless chains of inclusion ("turtles all the way down") and the circular chains of inclusion needed for self-reference. Barwise and Etchemendy [113] provide a readable

---

[6] Abbott [109] attributes this quotation to Johnson [110]; I am unable to find it in that volume.

account of circularity; Hofstadter [115, 116] is the maestro of self-reference and 'strange loops'.

The more recent form of this approach is found in category-theoretic coalgebras with their corecursion; these are considered advanced mathematical topics.

3.3.2 Advanced Dynamical Systems Theories

Dynamical systems model many processes in physics (Sect. 2.2). In classical approaches, the behaviour of the system depends deterministically on the instantaneous internal state of the system. However, living and life-like systems have properties that do not satisfy these constraints. There are more advanced branches of dynamical systems theory that cover these aspects.

Openness. Life is an open system. Non-autonomous differential equations allow the modelling of environmental inputs through a time-dependent function (driven simple harmonic motion is a classic example).

Stochasticity. Living systems are variable, messy, and stochastic. Stochastic DEs allow modelling non-determinism in the form of noise. Not all variation in living systems is noise or error to be reduced: variation is the very driving force of evolution.

Memory and adaptation. Living systems have memory and adapt to experience and circumstance. Time-delay DEs and integro-differential equations allow modelling memory, or history of past events. Fractional order DEs allow modelling of long-term memory, phase transitions, and fractal structure and dynamics in complex systems [117].

Hybrid. Living systems tend to have components and processes that are best modelled by a combination of discrete (for the solid) and continuous (for the fluid) spatial components, and also a combination of discrete and continuous processes (both stepping and flowing, for example, a bouncing ball). This requires a close integration of two modelling approaches, for example, using hybrid dynamical systems. Further generalisation could include encompassing discrete-symbolic, probabilistic, and dynamical systems oriented views [24].

Growth. A growing system involves a change in the dimensionality of the modelled state space (Sect. 3.9): new dimensions mean new equations (for example, the production of new molecules in a chemical reaction network, or new species in a food web). This requires changing the model, which potentially requires self-reflection.

Jaeger [118] discusses what he dubs *wild systems* (such as the brain): high dimensional heterogeneous open systems driven by fast stochastic inputs, with non-stationary dynamical laws changing as a result of restructuring, evolution, and growth, and says that these "are wilder than today's dynamical systems theory can handle".

3.4 Simulation and In Silico Models

One approach to tackling such wild systems is to use computational, rather than purely mathematical, models and analyses.

3.4.1 Computational Simulation of Models

Simulating any system, evolved or engineered, needs to be done in a principled manner, based on the specific research questions or engineering goals, in order to develop a simulation that is fit-for-purpose [119]. Depending on the goals, simulations can be performed at different levels of abstraction, from high-level concepts and relationships to low-level details of specific functions. Irrespective of the level of abstraction, simulations can require considerable computational power, due to the complexity and scale of biological systems and engineered systems (for example, highly instrumented cities).

The specific models chosen to be the basis of the simulation need to match the relevant underlying structures of (the simulated aspects of) system under study. In addition to the physical and biological aspects needed, full models will also need to include the relevant computational and self-referential aspects. This may preclude certain classical approaches. For example, Danchin [3] notes that there is a "trend in systems biology, in which recursivity and information replace the usual concepts of differential equations, feedback and feedforward loops", and that "many of the models used in systems biology rely on hypotheses (continuous differential equations in particular) that are often too crude to offer a realistic representation of the cell."

3.4.2 Agent-Based Models (ABMs) and Simulation

ABM is in some sense the antithesis of dynamical systems modelling. Instead of modelling a global state space, and a system history as a trajectory through that space, ABM starts with individual conceptual agents, with potentially complex internal states and behaviours, that sense, move, and interact in some environment. Simulation then animates the model, allowing histories to be determined. This leads to a more experimental approach to model analysis.

ABMs can naturally incorporate growth, since spawning a new agent, as a result of some growth or reproductive behaviour, increases the number of dimensions in the model. Much work on computational morphogenesis (Sect. 2.5.2) uses ABMs. It is important to take note of biological processes when designing growth processes, and not simply use a default clocked approach. For example, there are synchronisation changes across different domains during biological development [120].

3.4.3 Computational Self-reference

As noted in Sects. 3.2.4, 3.3.1, and 3.7, living systems are self-referential. This self-reference is realised in the computational aspects of the systems. Computers can perform self-reference via reflection [19]: code can 'see itself', refer to itself, run itself, and modify itself. Indeed, modern computer architectures are designed to separate code and data, specifically to protect against (accidental) self-modification.

Classic ABMs are typically not self-referential at the code level. Typically, these are implemented in some form of object-oriented language, with fixed classes determining the possibilities. So the system is limited to a combinatoric assembly of pre-existing structures. Even if an agent can reflect on its own inner state, it does not typically modify its own code to produce new modes of sensing, locomotion, or reasoning, for example. It can be argued that self-referential ability is necessary for open-ended behaviours [9, 121]. Automata chemistries (where the agents are strings of assembly language code) are one suitable medium for building models that incorporate self-reference [122], and these can exhibit a form of semantic closure [123].

3.4.4 Computational Limitations

Landauer [124] notes a further consequence of the fact that computing is subject to physical limitations; these limit not just our computational capabilities, but also our theories: they constrain what we can calculate, and hence constrain the complexity of feasible theories. Landauer is concerned with physical laws, but this argument also applies to what we can calculate about (our models and theories of) living systems, which are large, messy, and complex, not readily amenable to simplifying assumptions. We may simply not be able to build and explore accurate and precise models of living systems at all the relevant scales: we may be restricted to only "a crude look at the whole" [125].

3.5 Scale, Complexity, and Emergence

Living systems and living technologies are large, complex, complicated,[7] and messy [127, Chap. 2]. They self-organise around their complexity through emergence and hierarchical structures. Such multi-scale complexity and emergence introduce their own challenges to understanding (see also Sect. 3.2.5).

Living systems have large numbers of certain components: trillions of cells making a human body; trillions more in the gut biome; billions of DNA bases; millions and

---

[7] The distinction between complexity and complication is not a sharp distinction, but may be thought of thus: Complexity is associated with dynamic, bottom-up self-organisation, as in complexity science, while complication is associated with top-down organisational structure, as often in engineering; systems with both features have been dubbed 'wicked' systems [126].

more in population sizes.[8] These huge numbers exhibit emergent properties: "more is different" [129]. For large numbers, macroscopic system approximations can often be made, although assumptions that all components of a certain type are 'identical' need to be treated with caution.

Contrariwise, other components, although crucial, are small in number. A cell has one instance of the DNA molecule, split into a few chromosomes, and small numbers of other macromolecules (while also having a huge number of smaller active molecules such as water). An organism has one or two instances of its major organs. For small numbers, microscopic system detailed investigations can often be made, although as open systems with context and feedback.

And then there are the intermediate scale mesoscopic system properties, where there are not enough particles for macroscopic averaging, but too many for feasible microscopic small number particle analysis. This domain can also be considered a length scale: intermediate between the atomic/molecular nanoscale and the everyday object macroscale, the mesoscale is where Brownian motion dominates [130].

Bains [131] describes the need to move "beyond the toy domain".[9] The very complexity and variation, in systems and their environments, the range of scales, are all key components and should not be simplified away until all that is left is a 'toy'.

There are many forms of emergence in the computational domain [132], but the one of most interest here is the idea of a Virtual Machine: an emulation of one computer (architecture) that runs on top of another (virtual or real) machine. A VM is a computational way of 'hiding' the computing substrate: for example, one cannot tell whether a given VM is running on a Windows, Apple, Linux, or other platforms, or indeed, on another VM.

VMs are not restricted to the digital computing realm; they are also advocated as a way of structuring cognitive processing and consciousness in the brain [133]. Even ignoring such cognitive levels, the Reservoir Computing model can be considered as a VM for *in materio* computing [57]: a single model that can run on a wide range of physical embodiments, including soft robotic bodies [134].

This idea meshes with a hierarchical view of the structure of living systems, thinking of each level of life (Sect. 2.1) as 'running' on, or an emergent process of, a physical machine provided by the lower level.

3.6 Far-From-Equilibrium Thermodynamics

Equilibrium thermodynamics is a fundamental core part of physics. However, living systems are far from equilibrium. As with the development of non-linear dynamical

---

[8] The minimum viable population of a species has been estimated at around 3500–5000 individuals [128], which should be contrasted with the tens or hundreds making up a typical evolutionary algorithm 'population'.

[9] Bains' argument is in the context of Origin of Life, but is also relevant to the study of life itself.

systems through three rough stages (linear, small perturbations, fully non-linear, Sect. 2.2.3), thermodynamics has a similar range:

Equilibrium systems. These are typically closed or isolated systems,[10] moving to equilibrium, a state of maximum entropy. They can use free energy to do work.

Non-equilibrium systems. Such systems can be analysed as small perturbations from equilibrium.

Far-from-equilibrium systems. These are open systems, maintained (or self-maintaining) in a far-from-equilibrium state by using a flux of free energy and/or material in, and entropy and waste material out. They are driven by dissipative systems that live on and exploit energy gradients.

In such systems, equilibrium concepts such as free energy, temperature, and entropy are no longer so well-defined. This means that equilibrium thermodynamics intuitions can be misleading when reasoning about living systems. For example, complex structure and information can accumulate by exploiting noise through the use of ratchet mechanisms [135, 136].

3.7   Embodiment v. Virtual Physics

Everything is physical. Information, computing, and biological processes are embodied in structured physical material. Physicality provides the substance and constraints.

This leads to the question: can 'virtual life' exist? The Artificial Life community does much work *in silico*: is this a mere simulation, or can it ever be really alive? Can artificial chemistries [137, 138] and artificial physics support artificial (but still real) life? There is evidence that artificial resource limits can be exploited by simulated systems; for example, more diversity can occur in artificially evolving systems with 'mass' conservation [49, 139] and 'energy' constraints [140]. In a virtual system, there is a need to minimise the hard-wired 'physics', and allow as much of the simulated system as possible to evolve, to be 'soft' [141].

Can there be 'virtual embodiment'? It may be that embodiment is a property "of any suitably complex system engaged in a complex intertwined feedback relationship with its suitably complex environment" [61].

Rosen [142] disagrees. He builds a model where "life is closed to efficient cause", where every aspect is entailed by another aspect; in particular, he considers metabolism f, repair of metabolism Φ, and repair of metabolism b. Kercel [143], who provides an informative summary of Rosen's position, summarises this as $\Phi \vdash f \vdash b \vdash \Phi$. This is an inherently circular, self-referential definition (Sect. 3.4.3), similar to that of autopoiesis (Sect. 3.2.4).

Rosen argues that his model entails a need for physics to provide the situation that *grounds*, or gives meaning to, the ambiguous circularity and that this meaning can-

---

[10] In a closed system, matter cannot move in or out, but energy can, for example, in a closed system in a constant temperature heat bath. In an isolated system, energy is likewise banned from moving in or out and is hence conserved.

not be provided in a simulation with its ungrounded virtual physics. He also argues that it demonstrates the inadequacy of mere mechanistic models, which do not have this closed loop of causation, this property of self-definition. However, a simulation running on a universal computer is not a 'mere' machine (Sect. 3.9): universal computation supports self-reference and circularity, unlike mechanical devices. Hence, a virtual physics in a simulation sufficient to support self-reference, strange loops, and self-modifying code may we be sufficient to support (virtual) life.

Danchin [33] argues (Sect. 2.7.1) that the key to life is the symbolic relationships. These could be based on other physico-chemical substrates [144]; could they even include virtual *in silico* substrates, contrary to Rosen here? It seems plausible: life evolves the abstract symbolic control layer, a virtual machine; the ambiguity inherent in self-reference is resolved not by physical grounding, but by the particular, if arbitrary, symbolic representation chosen. Indeed, the demonstration of this arbitrariness can be used to distinguish intrinsic computing from non-computational processes in organisms [65]. It is not just symbols that can be implemented in a variety of ways: whole conceptual models can. As mentioned in Sect. 2.5.4, populations and networks are abstract models realised in a host of different manners. Danchin argues the symbols gain meaning only in context: "A cell can be defined as a machine that puts the genetic program into operation *according to the data provided by its environment*" [33, p. 270]. That meaning and context could be provided by a virtual environment.

However, Rosen's arguments are subtle and difficult and deserve further attention: if he is wrong, just where and why is he wrong? But if he is right, so much the worse for virtual ALife.

3.8   Metamaterials

Living systems are embodied in structured physical material. The structure is crucial to the material's ability to support complex information storage and processes. In biology, these materials are typically complex information-bearing polymers (DNA, RNA, and proteins).

Metamaterials are materials that have been highly engineered to have functional properties not present in ordinary materials. These can be electrical, optical, mechanical, or other properties. These properties can be programmable [145].

Metamaterials can be engineered to have computational properties. Classical computing takes place in what can be considered as computational metamaterials: silicon chips. Other metamaterials can be engineered to support other forms of classical computing, for example, mechanically realised boolean logic [146]. However, classical computing is at odds with the more 'natural' functions of materials [147]. Metamaterials can also be engineered to have unconventional computing properties [148], for example, as the substrate for *in materio* reservoir computing [58] and more general neuromorphic computing [149], for optical analogue computing [150], and as fluidic patterned controllers for soft robotics [151].

With suitable engineering, metamaterials can have simultaneous computational and physical functional properties: 'smart functional matter'. For example, oscillating MEMS (micro-electromechanical system) beams can be used as accelerometers and can also be used as reservoir computers [152]; combining these functionalities gives a device that can sense acceleration and process its sensing via reservoir computing in the same metamaterial [153]. Soft condensed matter might provide a stepping stone to more biological metamaterials [154].

We can then think of natural living substrates as highly *evolved* metamaterials, with both functional and computational properties. Indeed, DNA can be used as an engineered evolved metamaterial for DNA tiling [155] and DNA origami [156, 157]: its evolved information-bearing properties are engineered into self-assembling molecular-scale physical structures. In order to discover appropriate metamaterials for unconventional forms of living computation, it may be necessary to co-design the computational model and its supporting metamaterial [158].

So artificial living systems will be embodied in highly engineered computational and functional metamaterials. Virtual life, if it is possible (Sect. 3.7), will need appropriate virtual metamaterials for its embodiment.

3.9   Meta-Dynamics

The *machine metaphor*, that living systems are, or can be considered to be, *machines*, is a prevalent one. Some argue about the cultural meaning of the metaphor itself [159]. Others argue against its use in biology, for example, Woese [160] says:

> Let's stop looking at the organism purely as a molecular machine. The machine metaphor certainly provides insights, but these come at the price of overlooking much of what biology is. Machines are not made of parts that continually turn over, renew. The organism is. Machines are stable and accurate because they are designed and built to be so. The stability of an organism lies in resilience, the homeostatic capacity to reestablish itself.

I suggest that, rather than drawing a categorical distinction between machines and organisms, thereby running the risk of imputing the latter with some sort of *élan vital*, we should consider a spectrum, with 'mere' machines at one end, and living machines at the other.

Rao [161, p. 144] links the difference between 'mere' machine and organic processes as the nature of the changes they undergo:

> Where change in the machine metaphor is a process of stepwise re-engineering, in the other, more organic metaphors, change is a process of generative growth, ontogeny and self-organization.

So the spectrum of interest captures the forms of a structural change the system undergoes: it is along the dimension of the system's meta-dynamics (the dynamics of dynamics).

To reiterate, a dynamical system (Sect. 2.2.3) has a *state variable* (a value, typically a scalar or vector quantity) and a *state space* encompassing all possible values the state variable can take; a system's *dynamics* is a trajectory through the state space: the sequence of values its state variable takes through time. For meta-dynamics, the variable is itself a state space, and the (meta)state space is the set of all these values (all these state spaces); a (meta)trajectory moves through this meta-state space: the sequence of state spaces passed through with time. If the meta-dynamics is on a much longer timescale than the underlying dynamics, the two processes can be considered separately to some degree.

Physics uses dynamical systems as a modelling approach (Sect. 2.2.3), typically assuming a pre-defined, fixed dimensional state space. Computation can be viewed in terms of dynamical systems too [162], and the state space itself changes as variables go in and out of scope. These changes are not typically thought of in (meta)dynamical systems terms; work that does consider this aspect to some degree includes $(DS)^2$, 'dynamical systems with a dynamical structure' [163, 164], discrete dynamical systems automata [165], and the dynamics of networks (where the nodes represent variables, so new nodes represent new variables in a changing state space) such as in chemical organisation theory (COT):

> Complex dynamical reaction networks consisting of many components that interact and produce each other are difficult to understand, especially, when new component types may appear and present component types may vanish completely [166].

Living, growing systems naturally have a state space of increasing dimension and naturally have a meta-dynamics.

The spectrum from mere machine to living system can then be characterised by the trajectories of its meta-dynamics:

Mere machines, mechanisms, have a dynamics (behaviour/motion), but a very limited and mostly extrinsic meta-dynamics. They do wear and break, but any maintenance, repairs, upgrades, remodelling, debugging, etc. are extrinsic changes, happening discretely. Their functions, their trajectories, are typically cyclic, involving resetting to a given earlier state in the fixed state space, for example, thermodynamic work cycles and factory machines producing widget after widget. They have essentially no meta-dynamics.

Computational machines, computers, are more complex. Due to computational incompleteness and uncomputability, their dynamics can be unpredictable, only discoverable through executing the relevant program. Even if individual algorithms are closed during execution, overall a computer is an open dynamical system [167], taking user inputs, and providing a stream of outputs, throughout its operation. This open operation of a computer does not typically involve a reset to an earlier state: it takes input, stores data, learns, adapts, and is upgraded, all of which are a form of both intrinsic and extrinsic meta-dynamics, as its state space changes. But in principle computers can return (or rather, be returned) to earlier check pointed states; their trajectories can contain occasional cycles: rolling back a transaction, switched off and on again, factory reset. They have weak meta-dynamics.

Living machines, organisms, are characterised by a complex and rich intrinsic meta-dynamics: they continuously self-maintain, develop, grow, mature, behave, learn, etc.

Much of this meta-dynamics is intrinsic (although to a degree also extrinsic, due to interaction with a complex of other organisms and a physical environment). Living systems typically cannot be reset to an earlier state: the change is irreversible, and the trajectories are not cyclic. They have highly non-trivial intrinsic meta-dynamics.

4   Zoetic Engineering

So to progress Zoetic Science, the science of living systems, evolved, engineered, or possibly virtual, requires specific advancements in dynamical and meta-dynamical open systems, simulation, far-from-equilibrium thermodynamics, and metamaterials to cope with self-reference and self-production, multi-scale modelling, embodied computing, and growth. There will be more areas needing advancement, too. How to make progress with all these seemingly disparate branches of science? And how to ensure that the advances in the sub-disciplines are the right ones for the overall domain, and do not simply wander off into areas that are tractable but irrelevant?

We can take a lesson from the development of classical thermodynamics. Today it is core physics, a scientific discipline. However, historically during the Industrial Revolution, it was developed in response to an *engineering* need: to understand the fundamental limits to the efficiency of steam engines.

Engineering, in a nutshell, is making the world *as we want it to be*,[11] by building artefacts of increasing sophistication and capability. We could develop a new discipline specifically of Zoetic Engineering, developing living technologies of increasing sophistication, and use that to kick-start a grounded discipline of Zoetic Science. This would ideally result in a tight feedback loop of increasingly sophisticated artefacts, requiring and grounding increasingly sophisticated scientific advances, enabling further progress in the engineering domain, and so on.

Today, the technology that *constructs* an artefact is separate from the technology *of* the artefact, even biomimetic artefacts. Tomorrow, the technology is *both* the artefact *and* its construction and maintenance, all in one: a 'living' technology. With the domain focus on self-production through mesoscale self-organisation, self-assembly, and growth, an initial application of Zoetic Engineering might be a novel manufacturing process: self-producing artefacts.

Two of the example domains noted above (Sect. 2.7.2), swarm robotics and morphogenetic architecture, might provide fruitful starting points. The living technology would be an open system, responsive to and adapting to inputs during its lifetime. For architecture, the gardening metaphor is appropriate, providing inputs to guide and shape growth; for swarms, a shepherding metaphor might also find use. The artefact does not have to be entirely living, it can benefit from a hybrid approach. Growth could exploit classical building (such as scaffolding or trellis for support, and inorganic pipes and wires). It could also exploit a complementary living con-

---

[11] One might argue that most engineering projects fail in this regard, due to our inability, or lack of desire, to anticipate and mitigate for all the related unwanted consequences.

struction technology based on assemblers of termite-like swarms; swarms can also interact with the physical environment in a complex manner [168]. Then 'gardening' (planting, training, pruning [169], landscaping) and 'shepherding' tailor the growth context [84, 111].

5 Conclusions

Living systems are machines, but are not 'mere' machines. They are highly evolved or engineered, self-referential, computational, adaptive, stochastic, far-from-equilibrium machines, embodied in smart functional metamaterials, with non-trivial meta-dynamics: *meta-machines*,[12] if you like.

Understanding and engineering such zoetic systems require a new synthesis of biology, computing, and physics, where all three disciplines are extended and moved out of their classical comfort zones. This overview has omitted any detailed discussion of several other linked disciplines which are also part of the overall synthesis, including abiogenesis, the origins of life [172]; biochemistry (all the non-DNA/protein but nonetheless complex biochemicals in cells), chemistry (including the self-assembly of complexity through constrained combinatorics) and chemical engineering; computationalism (the view that the brain is an information-processing organ, a computer, implying that a computer could be a brain, could think) [74]; cybernetics and control systems, including second-order (recursive, self-referential) cybernetics; the ethics of living artefacts; healthy flourishing systems *v.* diseased systems, including ecological concepts from symbiosis and parasitism; neuromorphic computing; probability and statistics; and more.

Zoetic Science could be grounded by Zoetic Engineering: the production of living, or life-like, technologies and artefacts. Where might this lead? I am reminded of the statement by James Burke in his TV series *Connections*, on the long complex history of new ideas and new technologies, that themselves lead to further ideas and technologies:

> we live in a situation we inherited, as a result of a long and complex series of events through history. At no time in the past could anybody have known that what they were doing then would end up like this now [173].

---

[12] The prefix 'meta' originally meant simply 'after', but came to the additional meaning of 'above' or 'transcending', and now also includes the meaning of 'change' or 'transformation', and more recently of 'self-reference' (as in meta-X is the X of X). The 'transcending' meaning comes from a misunderstanding of the derivation of the word 'metaphysics': the term as originally coined did not mean 'transcending physics', but rather as 'Aristotle's book after the one called Physics'.

'Metamaterials' are *changed* materials: changed by engineering in this case. 'Meta-dynamics' is the *self-referential* use: the dynamics of dynamics. Given the prevalence of the prefix in the topics of interest here, it would be nice to be able to bundle the concepts into the term 'metaphysics', but it has already been taken for that unrelated Aristotelian topic. (To add further confusion to naming, 'meta-dynamics' is also the name of a computational simulation technique [170, 171], although the original 2002 paper does not name it thus.)

Acknowledgements Some of the ideas and topics presented here grew from a collaboration with Rupert Soar, while co-organising an EPSRC-funded 'Big Ideas' workshop in 2019, and from the attendees and presenters at that workshop: Martyn Amos, Rachel Armstrong, Vassilis M. Charitopoulos, René Doursat, Sarah Harris, Tim Ireland, Andrew Jenkins, Veronika Kapsali, Natalio Krasnogor, Ottoline Leyser, Andy Lomas, Tomasz Liskiewicz, Richard James MacCowan, Alison McKay, Robin Ramphal, Robert Richardson, and Tia Shaker.

My further thanks to Rachel Armstrong, Leo Caves, Herbert Jaeger, Rupert Soar, Adam Stanton, Tom McLeish, and three anonymous reviewers, for perceptive and constructive comments on an earlier draft of this paper.

References

1. Dobzhansky, T.: Nothing in biology makes sense except in the light of evolution. Am. Biol. Teach. **35**(3), 125–129 (1973)
2. Bains, W.: What do we think life is? A simple illustration and its consequences. Int. J. Astrobiol. **13**(2), 101–111 (2014)
3. Danchin, A.: Bacteria as computers making computers. FEMS Microbiol. Rev. **33**(1), 3–26 (2009)
4. Langton, C.G.: Artificial life. In: Artificial Life: the Proceedings of an Interdisciplinary Workshop on the Synthesis and Simulation of Living Systems, pp. 1–47. Addison-Wesley (1988)
5. Keller, E.F.: Making Sense of Life: Explaining Biological Development with Models, Metaphors, and Machines. Harvard University Press (2002)
6. James, W.: Principles of Psychology, vol. I. Hentry Holt and Co. (1890)
7. Smith, J.M., Szathmáry, E.: The Major Transitions in Evolution. Oxford University Press (1995)
8. Szathmáry, E.: Toward major evolutionary transitions theory 2.0. PNAS **112**(33), 10104–10111 (2015)
9. Banzhaf, W., Baumgaertner, B., Beslon, G., Doursat, R., Foster, J.A., McMullin, B., de Melo, V.V., Miconi, T., Spector, L., Stepney, S., White, R.: Defining and simulating open-ended novelty: requirements, guidelines, and challenges. Theory Biosci. **135**(3), 131–161 (2016)
10. Stepney, S.: Modelling and measuring open-endedness. In: OEE4 Workshop, at ALife 2021, Prague, Czech Republic (online) (2021)
11. Beslon, G., Elena, S., Hogeweg, P., Schneider, D., Stepney, S.: Evolving living technologies—insights from the EvoEvo project. In: SSBSE 2018, Montpellier, France. LNCS, vol. 11036, pp. 46–62. Springer (2018)
12. Marshall, S.M., Mathis, C., Carrick, E., Keenan, G., Cooper, G.J.T., Graham, H., Craven, M., Gromski, P.S., Moore, D.G., Walker, S.I., Cronin, L.: Identifying molecules as biosignatures with assembly theory and mass spectrometry. Nat. Commun. **12**(1), 3033 (2021)
13. Schneider, E.D., Sagan, D.: Into the Cool: Energy Flow, Thermodynamics, and Life. University of Chicago Press (2005)
14. Russell, M.J.: The alkaline solution to the emergence of life: energy, entropy and early evolution. Acta. Biotheor. **55**(2), 133–179 (2007)
15. Campbell, D., Farmer, D., Crutchfield, J., Jen, E.: Experimental mathematics: the role of computation in nonlinear science. Commun. ACM **28**(4), 374–384 (1985)
16. Strogatz, S.H.: Nonlinear Dynamics and Chaos: With Applications to Physics, Biology, Chemistry, and Engineering, 2nd edn. Westview Press (2015)
17. Abraham, R.H., Shaw, C.D.: Dynamics: The Geometry of Behavior, 2nd edn. Addison-Wesley (1992)
18. Turing, A.M.: The chemical basis of morphogenesis. Philos. Trans. R. Soc. Lond. B **237**(641), 37–72

19. Maes, P.: Concepts and experiments in computational reflection. In: OOPSLA '87, pp. 147–155. ACM (1987)

20. Horsman, D., Stepney, S., Kendon, V.: The natural science of computation. Commun. ACM **60**, 31–34 (2017)

21. Wegner, P.: Why interaction is more powerful than algorithms. Commun. ACM **40**(5), 80–91 (1997)

22. Shannon, C.E.: Mathematical theory of the differential analyzer. J. Math. Phys. **20**(1–4), 337–354 (1941)

23. Broersma, H., Stepney, S., Wendin, G.: Computability and complexity of unconventional computing devices. In: Stepney et al. [148], pp. 185–229

24. Jaeger, H.: Toward a generalized theory comprising digital, neuromorphic, and unconventional computing. Neuromorph. Comput. Eng. **1**(1) (2021)

25. Goodwin, B.C.: How the Leopard Changed Its Spots: The Evolution of Complexity. Phoenix (1994)

26. Jun, S., Mulder, B.: Entropy-driven spatial organization of highly confined polymers: lessons for the bacterial chromosome. PNAS **103**(33), 12388–12393 (2006)

27. Savage, N.: Computer logic meets cell biology: how cell science is getting an upgrade. Nature **564**(7734), S1–S3 (2018)

28. Yan, K.-K., Fang, G., Bhardwaj, N., Alexander, R.P., Gerstein, M.: Comparing genomes to computer operating systems in terms of the topology and evolution of their regulatory control networks. PNAS **107**(20), 9186–9191 (2010)

29. Grozinger, L., Amos, M., Gorochowski, T.E., Carbonell, P., Oyarzún, D.A., Stoof, R., Fellermann, H., Zuliani, P., Tas, H., Goñi-Moreno, A.: Pathways to cellular supremacy in biocomputing. Nat. Commun. **10**(1), 5250 (2019)

30. Jones, N.D., Gomard, C.K., Sestoft, P.: Partial Evaluation and Automatic Program Generation. Prentice Hall (1993)

31. Mark Baldwin, J.: A new factor in evolution. Am. Nat. **30**(354), 441–451 (1896)

32. Adami, C.: Introduction to Artificial Life. Springer (1998)

33. Danchin, A.: The Delphic Boat: What Genomes Tell Us. Harvard University Press (2002)

34. Bennett, C.H.: Logical depth and physical complexity. In: Herken, R., (ed.), The Universal Turing Machine: A Half-Century Survey, pp. 227–257. Oxford University Press (1988)

35. Lindenmayer, A.: Mathematical models for cellular interactions in development I. Filaments with one-sided inputs. J. Theor. Biol. **18**(3), 280–299 (1968)

36. Lindenmayer, A.: Mathematical models for cellular interactions in development II. Simple and branching filaments with two-sided inputs. J. Theor. Biol. **18**(3), 300–315 (1968)

37. Prusinkiewicz, P., Lindenmayer, A.: The Algorithmic Beauty of Plants. Springer (1990)

38. Doursat, R., Sayama, H., Michel, O., (eds.), Morphogenetic Engineering: Towards Programmable Complex Systems. Springer (2012)

39. El Karoui, M., Hoyos-Flight, M., Fletcher, L.: Future trends in synthetic biology-a report. Front. Bioeng. Biotechnol. **7**, 175 (2019)

40. Grohens, T., Meyer, S., Beslon, G.: A genome-wide evolutionary simulation of the transcription-supercoiling coupling. In: ALife 2021. MIT Press (2021)

41. Holland, J.H., (ed.) Adaptation in Natural and Artificial Systems, (2nd edn, 1992). MIT Press (1975)

42. Banzhaf, W., Nordin, P., Keller, R.E., Francone, F.D.: Genetic Programming. Morgan Kaufmann, An Introduction (1998)

43. Kennedy, J., Eberhart, R.C.: Particle swarm optimization. In: ICNN'95, Perth, Australia, vol. 4, pp. 1942–1948. IEEE (1995)

44. Kennedy, J., Eberhart, R.C.: Swarm Intelligence. Morgan Kaufmann (2001)

45. Dorigo, M., Stützle, T.: Ant Colony Optimization. MIT Press (2004)

46. Flower, D.R., Timmis, J., (eds.): In Silico Immunology. Springer (2007)

47. McCulloch, W.S., Pitts, W.: A logical calculus of the ideas immanent in nervous activity. Bull. Math. Biophys. **5**(4), 115–133 (1943)

48. Schmidhuber, J.: Deep learning in neural networks: an overview. Neural Netw. **61**, 85–117 (2015)
49. Lones, M.A., Fuente, L.A., Turner, A.P., Caves, L.S.D., Stepney, S., Smith, S.L., Tyrrell, A.M.: Artificial biochemical networks: evolving dynamical systems to control dynamical systems. IEEE Trans. Evolut. Comput. **18**(2), 145–166 (2014)
50. Bonabeau, E.W., Dorigo, M., Theraulaz, G.: Swarm Intelligence: From Natural to Artificial Systems. Addison Wesley (1999)
51. Aranha, C., Villalón, C.L.C., Campelo, F., Dorigo, M., Ruiz, R., Sevaux, M., Sörensen, K., Stützle, T.: The elephant in the room. Metaphor-based metaheuristics, a call for action. Swarm Intell. **16**, 1–6 (2021)
52. Newborough, J., Stepney, S.: A generic framework for population-based algorithms, implemented on multiple fpgas. In: ICARIS 2005, Banff, Canada. LNCS, vol. 3627, pp. 43–55. Springer (2005)
53. Lehman, J., Stanley, K.O.: Exploiting open-endedness to solve problems through the search for novelty. In: ALife XI, Winchester, UK, pp. 329–336. MIT Press (2008)
54. Lehman, J., Stanley, K.O.: Abandoning objectives: evolution through the search for novelty alone. Evol. Comput. **19**(2), 189–223 (2011)
55. Lones, M.A., Turner, A.P., Fuente, L.A., Stepney, S., Caves, L.S.D., Tyrrell, A.M.: Biochemical connectionism. Natl. Comput. **12**(4), 453–472 (2013)
56. Landauer, R.: Information is physical. Phys. Today **44**(5), 23–29 (1991)
57. Dale, M., Miller, J.F., Stepney, S.: Reservoir computing as a model for *in materio* computing. In Andrew Adamatzky, editor, Advances in Unconventional Computing, vol 1, pp. 533–571. Springer, 2017
58. Dale, M., Miller, J.F., Stepney, S., Trefzer, M.: Reservoir computing in material substrates. In: Nakajima, K., Fischer, I., (eds.), Reservoir Computing: Theory, Physical Implementations and Applications, pp. 141–166. Springer (2021)
59. Clark, A.: Being There: Putting Brain. Oxford University Press, Body and World Together Again (1997)
60. Pfeifer, R., Bongard, J.C.: How the Body Shapes the Way We Think: A New View of Intelligence. MIT Press (2007)
61. Stepney, S.: Embodiment. In: Flower and Timmis [46], chapter 12, pp. 265–288
62. Müller, V.C., Hoffmann, M.: What is morphological computation? on how the body contributes to cognition and control. Artif. Life **23**(1), 1–24 (2017)
63. Steane, A.: Quantum computing. Rep. Prog. Phys. **61**(2), 117 (1998)
64. Horsman, C., Stepney, S., Wagner, R.C., Kendon, V.: When does a physical system compute? Proc. R. Soc. A **470**(2169), 20140182 (2014)
65. Horsman, D., Kendon, V., Stepney, S., Young, J.P.W.: Abstraction and representation in living organisms: when does a biological system compute? In: Dodig-Crnkovic, G., Giovagnoli, R., (eds.), Representation and Reality in Humans, Other Living Organisms and Intelligent Machines, pp. 91–116. Springer (2017)
66. Stepney, S., Kendon, V.: The representational entity in physical computing. Nat. Comput. **20**(2), 233–242 (2021)
67. Russell, B., Stepney, S.: Zermelo navigation and a speed limit to quantum information processing. Phys. Rev. A **90**, 012303 (2014)
68. Lloyd, S.: Ultimate physical limits to computation. Nature **406**(6799), 1047–1054 (2000)
69. Landauer, R.: Irreversibility and heat generation in the computing process. IBM J. Res. Dev. **5**(3), 183–191 (1961)
70. Bennett, C.H.: Notes on the history of reversible computation. IBM J. Res. Dev. **32**(1), 16–23 (1988)
71. Gould, S.J.: Wonderful Life: The Burgess Shale and the Nature of History. Hutchinson (1989)
72. Lehman, J., Clune, J., Misevic, D., Adami, C., Beaulieu, J., Bentley, P.J., Bernard, S., Belson, G., Bryson, D.M., Cheney, N., Cully, A., Donciuex, S., Dyer, F.C., Ellefsen, K.O., Feldt, R., Fischer, S., Forrest, S., Frénoy, A., Gagneé, C., Goff, L.L., Grabowski, L.M., Hodjat, B., Keller, L., Knibbe, C., Krcah, P., Lenski, R.E., Lipson, H., MacCurdy, R., Maestre, C.,

Miikkulainen, R., Mitri, S., Moriarty, D.E., Mouret, J.-B., Nguyen, A., Ofria, C., Parizeau, M., Parsons, D., Pennock, R.T., Punch, W.F., Ray, T.S., Schoenauer, M., Shulte, E., Sims, K., Stanley, K.O., Taddei, F., Tarapore, D., Thibault, S., Weimer, W., Watson, R., Yosinksi, J.: The surprising creativity of digital evolution: a collection of anecdotes from the evolutionary computation and artificial life research communities. Artif. Life **26**(2), 274–306 (2020)

73. Laughlin, R.B.: A Different Universe: Reinventing Physics from the Bottom Down. Basic Books (2005)
74. Scheutz, M., (ed.), Computationalism: New Directions. MIT Press (2002)
75. Cohen, I.R.: Tending Adam's Garden: Evolving the Cognitive Immune Self. Academic (2000)
76. Schranz, M., Umlauft, M., Sende, M., Elmenreich, W.: Swarm robotic behaviors and current applications. Front. Robot. AI **7** (2020)
77. Brooks, R.A., Flynn, A.M.: Fast, cheap and out of control: a robot invasion of the solar system. J. Br. Interplanet. Soc. **42**, 478–485 (1989)
78. Kelly, K.: Out of Control: The New Biology of Machines. 4th Estate (1994)
79. Slavkov, I., Carrillo-Zapata, D., Carranza, N., Diego, X., Jansson, F., Kaandorp, J., Hauert, S., Sharpe, J.: Morphogenesis in robot swarms. Sci. Robot. **3**(25) (2018)
80. Bredeche, N., Haasdijk, E., Prieto, A.: Embodied evolution in collective robotics: a review. Front. Robot. AI, 5, 2018
81. Witkowski, O., Ikegami, T.: How to make swarms open-ended? Evolving collective intelligence through a constricted exploration of adjacent possibles. Artif. Life **25**(2), 178–197 (2019)
82. Hensel, M., Menges, A., Weinstock, M. (eds.): Techniques and Technologies in Morphogenetic Design. Architectural Design, vol. 76, no 2 (2006)
83. Ireland, T., Garnier, S.: Architecture, space and information in constructions built by humans and social insects: a conceptual review. Philos. Trans. R. Soc. B **373**(1753) (2018)
84. Stepney, S., Diaconescu, A., Doursat, R., Giavitto, J.-L., Miller, J.F., Spicher, A.: Evolving, growing, and gardening cyber-physical systems. In: Armstrong, R., (ed.), Experimental Architecture: Designing the Unknown, pp. 89–101. Routledge (2019)
85. Armstrong, R.: Soft Living Architecture: An Alternative View of Bio-informed Practice. Bloomsbury (2018)
86. Ren, L., Li, B., Wang, K., Zhou, X., Song, Z., Ren, L., Liu, Q.: Plant-morphing strategies and plant-inspired soft actuators fabricated by biomimetic four-dimensional printing: a review. Front. Mater. **8** (2021). https://onlinelibrary.wiley.com/toc/15542769/2006/76/2
87. Li, S., Wang, K.W.: Plant-inspired adaptive structures and materials for morphing and actuation: a review. Bioinspirat. Biomimet. **12**(1), 011001 (2016)
88. Bartlett, S., Wong, M.L.: Defining lyfe in the universe: from three privileged functions to four pillars. Life **10**(4) (2020)
89. Lazebnik, Y.: Can a biologist fix a radio?—or, what I learned while studying apoptosis. Cancer Cell **2**, 179–182 (2002)
90. Müller, K.H.: Second-order science: the revolution of scientific structures. Echoraum (2016)
91. Evangelidis, B.: Space and time as relations: the theoretical approach of Leibniz. Philosophies **3**(2), 9 (2018)
92. Campbell, R.J., Bickhard, M.H.: Physicalism, emergence and downward causation. Axiomathes **21**(1), 33–56 (2011)
93. Caves, L., de Melo, A.T.: (Gardening) gardening: a relational framework for complex thinking about complex systems. In: Walsh and Stepney [174], pp. 149–196
94. Gilman, C.P.: Human Work. McClure, Phillips and Co., (1904)
95. Ingold, T.: An anthropologist looks at biology. Man **25**(2), 208–229 (1990)
96. Cassirer, E.: An Essay on Man: An Introduction to a Philosophy of Human Culture. Yale University Press (1944)
97. Rescher, N.: Process Metaphysics: An Introduction to Process Philosophy. SUNY Press (1996)
98. Bickhard, M.H., Campbell, D.T.: Emergence. In: Andersen, P.B., Emmeche, C., Finnemann, N.O., Christiansen, P.V., (eds.), Downward Causation, chapter 14, pp. 322–348. Aarhus University Press (2000)

99. Bickhard, M.H.: The interactivist model. Synthese **166**(3), 547–591 (2009)
100. Bickhard, M.H.: Some consequences (and enablings) of process metaphysics. Axiomathes **21**, 3–32 (2011)
101. Galton, A., Mizoguchi, R.: The water falls but the waterfall does not fall: new perspectives on objects, processes and events. Appl. Ontol. **4**(2), 71–107 (2009)
102. Meadows, D.H.: Thinking in Systems: A Primer. Earthscan (2008)
103. von Bertalanffy, L.: General System Theory: Foundations, Development, Applications. George Braziller (1968)
104. Alexander, C.: A city is not a tree. Architect. Forum **122**(1), 58–62 (1965)
105. DeLanda, M.: A Thousand Years of Nonlinear History. Zone Books (1997)
106. Varela, F.G., Maturana, H.R., Uribe, R.: Autopoiesis: the organization of living systems, its characterization and a model. Biosystems **5**(4) (1974)
107. Beer, R.D.: Autopoiesis and cognition in the game of life. Artif. Life **10**(3), 309–326 (2004)
108. McMullin, B.: Thirty years of computational autopoiesis: a review. Artif. Life **10**(3), 277–295 (2004)
109. Porter Abbott, H.: Unnarratable knowledge: the difficulty of understanding evolution by natural selection. In: Herman, D., (ed.), Narrative Theory and the Cognitive Sciences, pp. 143–162. CSLI (2003)
110. Johnson, S.: Emergence: The Connected Lives of Ants. Cities, and Software. Penguin, Brains (2001)
111. Miller, J.F.: The software garden. In: Walsh and Stepney [174], pp. 201–212
112. Irvine, A.D., Deutsch, H.: Russell's paradox. In: Zalta, E.N. (ed.) The Stanford Encyclopedia of Philosophy. Stanford University, Metaphysics Research Lab. Springer (2021)
113. Barwise, J., Etchemendy, J.: The Liar: An Essay on Truth and Circularity. Oxford University Press (1987)
114. Aczel, P.: Non-well-Founded Sets. CSLI (1988)
115. Hofstadter, D.R.: Gödel, Escher, Bach: An Eternal Golden Braid. Penguin (1979)
116. Hofstadter, D.R:. I am a Strange Loop. Basic Books (2007)
117. West, B.J.: Colloquium: fractional calculus view of complexity: a tutorial. Rev. Mod. Phys. **86**(4), 1169–1186 (2014)
118. Jaeger, H.: Today's dynamical systems are too simple: commentary to Tim van Gelder's "The dynamical hypothesis in cognitive science". Behav. Brain Sci. **21**(5), 643–644 (1998)
119. Stepney, S., Polack, F.A.C., Alden, K., Andrews, P.S., Bown, J.L., Droop, A., Greaves, R.B., Read, M., Sampson, A.T., Timmis, J., Winfield, A.F.T.: Engineering Simulations as Scientific Instruments: A Pattern Language. Springer (2018)
120. Foe, V.E.: Mitotic domains reveal early commitment of cells in Drosophila embryos. Development **107**(1), 1–22 (1989)
121. Stepney, S., Hoverd, T.: Reflecting on open-ended evolution. In: ECAL 2011, Paris, France, pp. 781–788. MIT Press (2011)
122. Hickinbotham, S., Stepney, S.: Bio-reflective architectures for evolutionary innovation. In: ALife 2016, Cancun, Mexico, pp. 192–199. MIT Press (2016)
123. Clark, E.B., Hickinbotham, S.J., Stepney, S.: Semantic closure demonstrated by the evolution of a universal constructor architecture in an artificial chemistry. J. R. Soc. Interface **14**, 20161033 (2017)
124. Landauer, R.: Wanted: a physically possible theory of physics. IEEE Spectr. **4**(9), 105–109 (1967)
125. Miller, J.H.: A Crude Look at the Whole: The Science of Complex Systems in Business, Life, and Society. Basic Books (2015)
126. Andersson, C., Törnberg, A., Törnberg, P.: Societal systems – complex or worse? Futures **63**, 145–157 (2014)
127. Hester, P.T., Adams, K.M.: Systemic Decision Making: Fundamentals for Addressing Problems and Messes, 2nd edn. Springer (2017)
128. Traill, L.W., Bradshaw, C.J.A., Brook, B.W.: Minimum viable population size: a meta-analysis of 30 years of published estimates. Biol. Conserv. **139**(1), 159–166 (2007)

129. Anderson, P.W.: More is different. Science **177**(4047), 393–396 (1972)
130. Haw, M.: Middle World: The Restless Heart of Matter and Life. Macmillan (2007)
131. Bains, W.: Getting beyond the toy domain. Meditations on David Deamer's "Assembling Life". Life **10**(2), 18 (2020)
132. Stepney, S.: Digital emergence. In: Gibb, S., Hendry, R.F., Lancaster, T., (eds.), Routledge Handbook of Emergence, pp. 329–338. Routledge (2019)
133. Sloman, A., Chrisley, R.: Virtual machines and consciousness. J. Conscious. Stud. **10**(4–5), 133–172 (2003)
134. Nakajima, K., Hauser, H., Kang, R., Guglielmino, E., Caldwell, D.G., Pfeifer, R.: A soft body as a reservoir: case studies in a dynamic model of octopus-inspired soft robotic arm. Front. Comput. Neurosci. **7**, 91 (2013)
135. Dasmahapatra, S., Werner, J., Zauner, K.-P.: Noise as a computational resource. Int. J. Unconvent. Comput. **2**(4), 305–319 (2006)
136. Hoffmann, P.M.: Life's Ratchet: How Molecular Machines Extract Order From Chaos. Basic Books (2012)
137. Banzhaf, W., Yamamoto, L.: Artificial Chemistries. MIT Press (2016)
138. Faulkner, P., Krastev, M., Sebald, A., Stepney, S.: Sub-symbolic artificial chemistries. In Stepney, S., Adamatzky, A., (eds.), Inspired by Nature: Essays Presented to Julian F. Miller on the Occasion of His 60th Birthday, pp. 287–322. Springer (2018)
139. Hickinbotham, S., Stepney, S.: Conservation of matter increases evolutionary activity. In: ECAL 2015, York, UK, pp. 98–105. MIT Press (2015)
140. Hoverd, T., Stepney, S.: Energy as a driver of diversity in open-ended evolution. In: ECAL 2011, Paris, France, pp. 356–363. MIT Press (2011)
141. Hickinbotham, S., Clark, E., Nellis, A., Stepney, S., Clarke, T., Young, P.: Maximising the adjacent possible in automata chemistries. Artif. Life J. **22**(1), 49–75 (2016)
142. Rosen, R.: Life Itself: A Comprehensive Enquiry into the Nature, Origin, and Fabrication of Life. Columbia University Press (1991)
143. Kercel, S.W.: Entailment of ambiguity. Chem. Biodivers. **4**(10), 2369–2385 (2007)
144. Bains, W.: Many chemistries could be used to build living systems. Astrobiology **4**(2), 137–167 (2004)
145. Tsilipakos, O., Tasolamprou, A.C., Pitilakis, A., Liu, F., Wang, X., Mirmoosa, M.S., Tzarouchis, D.C., Abadal, S., Taghvaee, H., Liaskos, C., Tsioliaridou, A., Georgiou, J., Cabellos-Aparicio, A., Alarcón, E., Ioannidis, S., Pitsillides, A., Akyildiz, I.F., Kantartzis, N.V., Economou, E.N., Soukoulis, C.M., Kafesaki, M., Tretyakov, S.: Toward intelligent metasurfaces: the progress from globally tunable metasurfaces to software-defined metasurfaces with an embedded network of controllers. Adv. Opt. Mater. **8**(17), 2000783 (2020)
146. Yasuda, H., Buskohl, P.R., Gillman, A., Murphey, T.D., Stepney, S., Vaia, R.A., Raney, J.R.: Mechanical computing. Nature **598**, 39–48 (2021)
147. Zauner, K.-P.: From prescriptive programming of solid-state devices to orchestrated self-organisation of informed matter. In: Banâtre, J.-P., Fradet, P., Giavitto, J.L., Michel, O., (eds.), Unconventional Programming Paradigms 2004. LNCS, vol. 3566, pp. 47–55. Springer (2005)
148. Stepney, S., Rasmussen, S., Amos, M., (eds.), Computational Matter. Springer (2018)
149. Kaspar, C., Ravoo, B.J., van der Wiel, W.G., Wegner, S.V., Pernice, W.H.P.: The rise of intelligent matter. Nature **594**(7863), 345–355 (2021)
150. Silva, A., Monticone, F., Castaldi, G., Galdi, V., Alù, A., Engheta, N.: Performing mathematical operations with metamaterials. Science **343**(6167), 160–163 (2014)
151. Garrad, M., Chen, H.-Y., Conn, A.T., Hauser, H., Rossiter, J.: Liquid metal logic for soft robotics. IEEE Robot. Autom. Lett. **6**(2), 4095–4102 (2021)
152. Dion, G., Mejaouri, S., Sylvestre, J.: Reservoir computing with a single delay-coupled non-linear mechanical oscillator. J. Appl. Phys. **124**(15), 152132 (2018)
153. Barazani, B., Dion, G., Morissette, J.-F., Beaudoin, L., Sylvestre, J.: Microfabricated neuroaccelerometer: integrating sensing and reservoir computing in MEMS. J. Microelectromech. Syst. **29**(3), 338–347 (2020)
154. Stepney, S.: The neglected pillar of material computation. Phys. D **237**(9), 1157–1164 (2008)

155. Evans, C.G., Winfree, E.: Physical principles for DNA tile self-assembly. Chem. Soc. Rev. **46**(12), 3808–3829 (2017)

156. Rothemund, P.W.K.: Folding DNA to create nanoscale shapes and patterns. Nature **440**(7082), 297–302 (2006)

157. Dey, S., Fan, C., Gothelf, K.V., Li, J., Lin, C., Liu, L., Liu, N., Nijenhuis, M.A.D., Saccà, B., Simmel, F.C., Yan, H., Zhan, P.: DNA origami. Nat. Rev. Methods Primers **1**(1), 1–24 (2021)

158. Stepney, S.: Co-designing the computational model and the computing substrate. In: UCNC 2019, Tokyo, Japan. LNCS, vol. 11493, pp. 5–14. Springer (2019)

159. Vaage, N.S.: Living machines: metaphors we live by. NanoEthics **14**(1), 57–70 (2020)

160. Woese, C.R.: A new biology for a new century. Microbiol. Mol. Biol. Rev. **68**(2), 173–186 (2004)

161. Rao, V.: Tempo: Timing, Tactics and Strategy in Narrative-Driven Decision-Making. Lightning Source (2011)

162. Stepney, S.: Nonclassical computation: a dynamical systems perspective. In: Rozenberg, G., Bäck, T., Kok, J.N., (eds.), Handbook of Natural Computing, chapter 59, vol. 4, pp. 1979–2025. Springer (2012)

163. Giavitto, J.-L., Michel, O.: MGS: A rule-based programming language for complex objects and collections. Electron. Notes Theor. Comput. Sci. **59**(4), 286–304 (2001)

164. Giavitto, J.-L., Michel, O., Cohen, J., Spicher, A.: Computations in space and space in computations, 137–152 (2005)

165. Nehaniv, C.L., Rhodes, J., Egri-Nagy, A., Dini, P., Morris, E.R., Horváth, G., Karimi, F., Schreckling, D., Schilstra, M.J.: Symmetry structure in discrete models of biochemical systems: natural subsystems and the weak control hierarchy in a new model of computation driven by interactions. Philos. Trans. R. Soc. A **373**(2046), 20140223 (2015)

166. Dittrich, P., di Fenizio, P.S.: Chemical organisation theory. Bull. Math. Biol. **69**(4), 1199–1231 (2007)

167. Stepney, S.: Computing with open dynamical systems. In: CogSIMA 2021, Tallin, Estonia (online), pp. 139–143. IEEE (2021)

168. Soar, R., Amador, G., Bardunias, P., Turner, J.S.: Moisture gradients form a vapor cycle within the viscous boundary layer as an organizing principle to worker termites. Insectes Soc. **66**(2), 193–209 (2019)

169. Prusinkiewicz, P., James, M., Měch, R.: Synthetic topiary. In: SIGGRAPH'94, pp. 351–358. ACM (1994)

170. Laio, A., Parrinello, M.: Escaping free-energy minima. PNAS **99**(20), 12562–12566 (2002)

171. Barducci, A., Bussi, G., Parrinello, M.: Well-tempered metadynamics: a smoothly converging and tunable free-energy method. Phys. Rev. Lett. **100**(2), 020603 (2008)

172. Walker, S.I., Packard, N., Cody, G.D.: Re-conceptualizing the origins of life. Philos. Trans. A 375(2109) (2017)

173. Burke, J.: Connections. Book Club Associates (1978)

174. Walsh, R., Stepney, S., (eds.), Narrating Complexity. Springer (2018)

STREAMLINE: A Simple, Transparent, End-To-End Automated Machine Learning Pipeline Facilitating Data Analysis and Algorithm Comparison

Ryan Urbanowicz, Robert Zhang, Yuhan Cui, and Pranshu Suri

Abstract Machine learning (ML) offers powerful methods for detecting and modeling associations often in data with large feature spaces and complex associations. Many useful tools/packages (e.g. scikit-learn) have been developed to make the various elements of data handling, processing, modeling, and interpretation accessible. However, it is not trivial for most investigators to assemble these elements into a rigorous, replicable, unbiased, and effective data analysis pipeline. Automated machine learning (AutoML) seeks to address these issues by simplifying the process of ML analysis for all. Here, we introduce STREAMLINE, a simple, transparent, end-to-end AutoML pipeline designed as a framework to easily conduct rigorous ML modeling and analysis (limited initially to binary classification). STREAMLINE is specifically designed to compare performance between datasets, ML algorithms, and other AutoML tools. It is unique among other autoML tools by offering a fully transparent and consistent baseline of comparison using a carefully designed series of pipeline elements including (1) exploratory analysis, (2) basic data cleaning, (3) cross validation partitioning, (4) data scaling and imputation, (5) filter-based feature importance estimation, (6) collective feature selection, (7) ML modeling with 'Optuna' hyperparameter optimization across 15 established algorithms (including less well-known Genetic Programming and rule-based ML), (8) evaluation across 16 classification metrics, (9) model feature importance estimation, (10) statistical significance comparisons, and (11) automatically exporting all results, plots, a PDF summary report, and models that can be easily applied to replication data.

R. Urbanowicz (✉)
Department of Computational Biomedicine, Cedars Sinai Medical Center, Los Angeles, CA, USA
e-mail: ryan.urbanowicz@cshs.org

R. Zhang · Y. Cui · P. Suri
University of Pennsylvania, Philadelphia, PA, USA

© The Author(s), under exclusive license to Springer Nature Singapore Pte Ltd. 2023
L. Trujillo et al. (eds.), *Genetic Programming Theory and Practice XIX*,
Genetic and Evolutionary Computation, https://doi.org/10.1007/978-981-19-8460-0_9

1 Introduction

Machine learning (ML) has become a cornerstone of modern data science with applications in countless research domains including biomedical informatics, a field synonymous with noisy, complex, heterogeneous, and often large-scale data (i.e. 'big-data') [32, 38, 42]. Surging interest in ML stems from its potential to train models that can be applied to make predictions as well as discover complex multivariate associations within increasingly large feature spaces (e.g. 'omics' data as well as integrated multi-omics data) [20]. Increased access to powerful computing resources has fueled the practicality of these endeavors [22]. As a result, a wealth of ML tools, packages, and other resources have been developed to facilitate the implementation of custom ML analyses. One popular and accessible example includes the scikit-learn library built with the Python programming language [36]. Packages such as scikit-learn focus on facilitating the use of individual elements of an ML analysis pipeline (e.g. cross validation, feature selection, and ML modeling). However, 'how' these elements are brought together is generally left up to the practitioner leading to significant variability in how ML analyses are conducted, even when dealing with similar data. Guidelines for conducting ML analysis largely exist as a community knowledge pool of individual 'potential pitfalls' and 'best practices' [24–26, 33, 39, 41, 44, 52]. Most ML research tends to focus on improving or adapting individual methods for a given data type, task, or domain of application. Surprisingly few works have focused on how to effectively and appropriately assemble an ML pipeline in its entirety or provide accessible, easy-to-use examples of how to do so. For those coming from outside domains of expertise, it can be daunting to know where to start.

In recent years, automated machine learning (i.e. AutoML) has emerged as a field of research, producing a number of strategies and tools aiming to facilitate and optimize the process of conducting machine learning data analyses [27]. Specific AutoML tools differ based on (1) which elements of an ML analysis pipeline they automate, e.g. feature selection, model selection, and/or hyperparameter optimization, (2) whether they seek to automate the design of the pipeline itself, i.e. what elements to include and in what order (e.g. TPOT which conducts pipeline optimization using genetic programming [35]), and (3) the types of data or application domains to which they were designed to be applied, e.g. tabular data versus images, discrete versus quantitative outcomes, etc. [19, 43, 53]. What makes most AutoML tools similar is that they focus on returning the single or subset of best results to the user. From the perspective of ML modeling, this constitutes returning the model yielding the best performance based on a target evaluation metric. The advantage of AutoML tools to date includes greater accessibility, reduced tedium and researcher time in developing ML analysis pipelines, and offering potentially better optimization of modeling in contrast with manually conducted analyses. However, AutoML tools can be extremely computationally expensive and they still cannot guarantee optimal performance, or even necessarily better performance than a pipeline manually developed by an expert [23, 43]. Furthermore, AutoML tools are each inherently limited by the specific algorithms they implement and automate, and they generally do little to educate or actively engage users with respect to the process. This could potentially create new opportunities for ML misuse when practitioners are less engaged in the design analyses, particularly with respect to potential sources of bias or assumptions introduced by either the study design/data source or the specific ML analyses conducted.

As such, there are many opportunities for future AutoML to answer questions such as the following: (1) Can automated ML pipeline assembly reliably give comparable or better performance than analyses conducted by experts? (2) What are the specific strengths and weaknesses of different ML algorithms in tasks with different data types, data sizes, and underlying problem complexity? (3) Can we engineer AutoML tools to be smarter? Either learn from past analysis experience of the tool to conduct analyses more efficiently and effectively in subsequent runs (e.g. PennAI [30]), or better constrain open ended AutoML search using human expertise on best practices and the use of specific algorithms? (4) Is there some subset of ML modeling algorithms that could be reliably applied more efficiently that represents a well-balanced cross section of algorithm strengths and weaknesses rather than applying all available algorithms? (5) How do we optimize ML interpretability? (6) How do we address covariates in ML analyses? (7) What is the best way to construct an ML analysis pipeline in different contexts? and (8) What aspects of a complete ML analysis pipeline can be reliably automated versus which are best conducted using domain expertise?

With these questions in mind, we developed a simple, transparent, end-to-end, automated machine learning pipeline (STREAMLINE) that focuses exclusively on binary classification in tabular data in this initial implementation. Unlike most other AutoML tools, STREAMLINE is designed as a framework to rigorously apply and compare a variety of ML modeling algorithms in a carefully designed and standardized manner. STREAMLINE adopts a fixed series of purposefully selected ML analysis pipeline elements in line with data science best practices. It seeks to automate any domain-generalizable elements of an ML analysis pipeline including exploratory analysis, basic data cleaning, cross validation (CV) partitioning, data scaling, missing value imputation, pre-modeling feature importance estimation, feature selection, ML modeling (including hyperparameter optimization within 15 scikit-learn [24] compatible ML algorithms that can be applied to binary classification), evaluation across 16 classification metrics, model feature importance estimation, generating and organizing all models, exporting publication-ready plots and results, statistical significance comparisons across ML algorithms and analyzed datasets, generation of a summary report of settings and key results, and easy application and evaluation of all trained models to replication data. Of the 15 modeling algorithms currently included in STREAMLINE, 11 are popular, well-established ML algorithms, 3 are rule-based evolutionary machine learning (RBML) algorithm implementations currently being developed by our group (i.e. eLCS, XCS, and ExSTraCS), and the last is a scikit-learn compatible implementation of genetic programming (GP) adapted for symbolic classification. We included the RBML algorithms to demonstrate how STREAMLINE can be used as a framework to test and compare new algorithms easily to other widely known standards. Further, the GP algorithm was included in part due to the relative ease of directly interpreting the models it produces in contrast with most other ML modeling algorithms, as well as to highlight GP as an accessible approach to modeling within the broader machine learning community.

The design of STREAMLINE focused on (1) overall automation, (2) avoiding or allowing easier detection of bias, (3) optimizing modeling performance, (4) ensuring reproducibility, (5) capturing complex associations in data (e.g. feature interactions), (6) enhancing interpretability of output, (7) offering different use modes for varying levels of user computing experience, (8) facilitating pipeline output re-use to conduct further analyses, and (9) making it easier to extend STREAMLINE in the future to include other algorithms or automated pipeline elements. Overall, the purpose of STREAMLINE is not to claim a 'best way' to conduct an ML or AutoML analysis, but rather to serve as a baseline framework with which to easily conduct a straightforward but rigorous ML modeling analysis over one or more target datasets, as well as to facilitate interrogation of the questions presented earlier.

In the following sections, we detail the automated elements of the STREAMLINE AutoML tool and provide an overview of (1) how it can be applied, (2) assumptions for use, and (3) output. Next, we include a demonstration of STREAMLINE applied to a well-known benchmark dataset from the UCI repository [21], as well as a variety of simulated genomics datasets with distinct underlying patterns of associations, and multiplexer ML benchmark datasets. Lastly, we discuss future directions for STREAMLINE as a tool to support data analysis, ML algorithm, and AutoML tool development and encourage a better understanding of pipeline assembly for conducting effective and reliable ML analysis.

2   Methods

In this section, we begin by detailing the automated elements of the STREAMLINE AutoML tool, discussing how it can be used, the assumptions it makes, and the output it produces. Then we will describe the datasets employed to test and demonstrate the efficacy of STREAMLINE. STREAMLINE is currently available at the following GitHub repository: https://github.com/UrbsLab/STREAMLINE [16]. The code version used in presented analyses was Beta 0.2.5. We refer readers to the README of this repository for detailed instructions regarding (1) dataset formatting requirements, (2) installation instructions, (3) pipeline run parameters, and (4) different use modes.

2.1   STREAMLINE

Figure 1 provides a schematic overview of the elements included in the STREAMLINE tool. This schematic identifies key individual elements of the pipeline organized into 4 general stages: (1) data preparation, (2) feature importance estimation

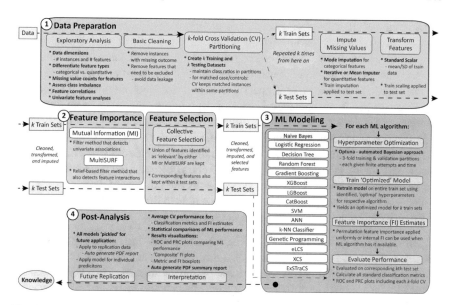

Fig. 1 STREAMLINE Schematic. For each target dataset, STREAMLINE proceeds through a fixed series of pipeline elements generalized into the 4 stages above. Arrows indicate pipeline flow through each, with arrows leaving stage 1 corresponding to those entering stage 2. Target data and subsequently generated training and testing datasets are indicated by yellow boxes. Specific elements of the pipeline are indicated by pink boxes and specific algorithms are indicated by white boxes. Key details of pipeline functionality are noted

and selection, (3) ML modeling, and (4) post-analysis. The overall arrangement of pipeline elements was designed to avoid bias and data leakage, as well as to maximize parallelizability when run on a computing cluster.

Below we detail the specific elements and reasoning behind their incorporation within each stage. Throughout this description, we provide select figures illustrating STREAMLINE output when applied to an example simulated genomics dataset with 100 single nucleotide polymorphism (SNP) features and 1600 instances including 2 heterogeneously independent 2-way predictive interactions described in Sect. 2.2. In practice, STREAMLINE offers recommended default run parameters that should work well for many analyses; however, we encourage users to explore available pipeline run parameters to see how they could be more flexibly adapted to different studies [16].

2.1.1 Data Preparation

This STREAMLINE stage focuses on helping users understand their data as well prepare the data for downstream modeling.

Data Input: STREAMLINE takes a folder including one or more properly formatted 'target' datasets as input for a single 'experiment', i.e. a complete run of the pipeline. This allows users to compare modeling performance across multiple datasets that might constitute scenarios such as (1) the original dataset with or without covariates, to evaluate their impact, (2) datasets that include different subsets of available features, to evaluate the impact of explicitly including or excluding those features, or (3) datasets constituting different underlying challenges for ML modeling. Key requirements for dataset formatting include numerical encoding of data values, a header of column labels, a unique and consistent identifier for missing data values, and having labeled data with a binary class outcome. All available instances can be included in these input data, or if a sufficient number of data instances are available, a random portion should ideally be withheld for replication analysis using STREAMLINE's 'apply-model' phase in post-analysis.

Exploratory Analysis: STREAMLINE summarizes each 'target' dataset using summary statistics and plots including (1) feature, instance, and missing value counts, (2) class balance, (3) feature correlation, and (4) univariate analyses. STREAMLINE automatically attempts to differentiate features that should be treated either as categorical or quantitative based on a (user-defined) cutoff of unique value counts. Users can also manually specify which features are to be treated as categorical. Depending on this distinction, appropriate univariate statistical tests and plots will be generated for each feature examining its relationship with outcome across the entire 'target' dataset. Notably, in this first release of STREAMLINE, the distinction of categorical versus quantitative features is only utilized in the first two stages. Scikit-learn modeling will treat all features as if they were quantitative by default. Users that wish to ensure that categorical features are treated as such in modeling should employ one-hot-encoding prior to running STREAMLINE. This element will be automated in a future release; however, we currently leave this decision up to users.

Basic Cleaning: Data cleaning is a notoriously difficult element to automate and often requires many analysis-specific decisions. STREAMLINE data cleaning limits itself to removing instances that are missing an outcome value, as well as removing any features specified by the user that should be excluded (e.g. to prevent data leakage). STREAMLINE does not automatically remove outliers, as this is a decision best left to the user based on domain knowledge.

k-fold Cross Validation: STREAMLINE employs k-fold cross validation (CV) prior to completing feature transformation, imputation, feature selection, or modeling to avoid any data leakage where information from the testing set is gleaned prior to final model evaluation. Users can choose from 3 CV strategies: (1) stratified, to ensure class balance within each partition, (2) random, and (3) matched, which allows users to keep specific groups of instances together within partitions [28]. Matched CV partitioning allows users to properly analyze covariate-matched datasets to help control for covariate effects in ML analyses [31]. STREAMLINE uses 10-fold stratified CV by default. From here and through modeling, all described elements of STREAMLINE are conducted within each of the k-fold partitions.

Impute Missing Values: Missing data values present practitioners with difficult decisions. Removing instances or features with missing values has the obvious drawbacks of reduced sample size and possibly removing relevant features, respectively. Some algorithm implementations are designed to handle missing values as neutrally as possible, which is likely the safest option, i.e. least likely to introduce unwanted bias [54]. However, STREAMLINE utilizes a number of scikit-learn ML modeling algorithm implementations that cannot handle missing values. As such, we view imputation here as a 'necessary evil' that will afford us the opportunity to more easily compare a broader range of widely used ML modeling algorithms. STREAMLINE currently employs simple mode imputation for categorical variables to ensure the same set of possible categorical values are used to replace missing ones and employs either iterative imputation [18] or mean imputation as options for quantitative features. Iterative imputation considers the context of all features in the dataset, while both mode and mean imputation simply use information from the target feature. While iterative imputation is likely to perform better, it can be computationally expensive to run and take up a great deal of disk space to save the imputer object for future model applications when applied to large datasets. In STREAMLINE, imputation is first conducted in the training data partition, and then the same imputation decisions learned in training are applied directly to the testing partitions, or later to any replication data when using 'apply-model'. Imputation is conducted prior to data scaling as the imputation could influence the correct center and scale to use.

Transform Features: Feature transformation can play an important role in the ability of specific ML modeling algorithms to effectively detect and model different patterns of association [29]. Generally speaking, more sophisticated ML algorithms can still effectively model associations without employing feature transformation. One exception to this is feature scaling, where feature values across the dataset are scaled to have the properties of a standard normal distribution with a mean of zero and a standard deviation of one. While not critical for all algorithms (e.g. decision trees), algorithms such as support vector machines (SVM), k-nearest neighbors (KNN), logistic regression (LR), and artificial neural networks (ANN) require the data to be scaled to ensure effective training and/or proper interpretation of the resulting models, i.e. feature importance estimation. STREAMLINE employs scikit-learn's 'standard scalar' to all datasets, performed first on the training data partition, and then applying the same learned scalar to testing or downstream replication data.

2.1.2 Feature Importance and Selection

This STREAMLINE stage focuses on providing a pre-modeling estimation of feature importance as well as seeking to conservatively remove features with no indication of being informative. We organized this stage as being separate from the rest of data processing, to reflect how this is a phase that can be viewed as part of both processing and modeling. It is an opportunity to gain further data insight about patterns in the data prior to modeling, as well as a potentially critical bottleneck in analyses dealing with large feature spaces, with many possible ways to go about it.

Notably, this first release of STREAMLINE does nothing to remove perfectly or highly correlated features from the data. We suggest users remove all but one feature of any fully correlated feature sets and consider further removal of highly correlated features based on the needs of the problem at hand.

Feature Importance Estimation: Prior to modeling, it can be valuable to examine the estimated feature importance of features in the dataset. This can facilitate down-stream interpretation, by providing something to compare model feature importance estimates to. Feature importance estimation can be based on a variety of feature selection (FS) strategies. While FS can take place in parallel with ML modeling (i.e. wrapper-based or embedded FS [48]), this can be computationally expensive, difficult to automate, and can identify feature subsets that are biased by the modeling algorithms being applied. Therefore, this pipeline focuses on utilizing filter-based FS methods, which are typically fast and can be combined with any downstream ML method. Unfortunately, most filter-based FS methods are insensitive to complex feature interactions. Ultimately, we are concerned with detecting both simple and potentially complex associations. As such, STREAMLINE implements two FS algorithms, i.e. mutual information (MI) [37] and MultiSURF [50] (a Relief-based FS algorithm). MI is proficient at evaluating univariate associations between a feature and outcome, while MultiSURF has been demonstrated to be sensitive to not only univariate associations but also both 2- and 3-way feature interactions (even in the absence of univariate associations).

Feature Selection: A key factor in the scalability of any AutoML tool will be the number of instances and features in the target data. Feature selection is an element of an ML analysis pipeline that, in some cases, can be conducted directly by ML modeling algorithms. However, in general, we can expect ML modeling algorithms to perform best if irrelevant or redundant features have been removed first. A critical mistake in conducting feature selection prior to ML modeling is to accidentally remove relevant features using over-simplistic selection methods such as statistical tests or algorithms that only account for simple univariate or linear associations. Previous research has suggested that FS conducted by an ensemble of methodologies provides an effective approach to avoid accidentally removing relevant features. As such, STREAMLINE adopts 'collective feature selection' that calls for the use of more than one FS methodology in making the determination as to whether a given feature should be retained or removed [51]. If either MI or MultiSURF finds evidence that a feature may be informative, that feature is retained; otherwise, it is removed. Both algorithms estimate feature importance scores for every feature where a score larger than 0 is considered to be potentially informative (and the feature is retained). In practice, irrelevant features can have values > 0 due to random chance in feature values, but we are conservative in selecting features to avoid removing features with potentially small but meaningful effects. Users can set STREAMLINE to keep all features regardless of feature importance estimates or set a cap on the maximum number of features to allow in each dataset prior to modeling. If the latter is selected,

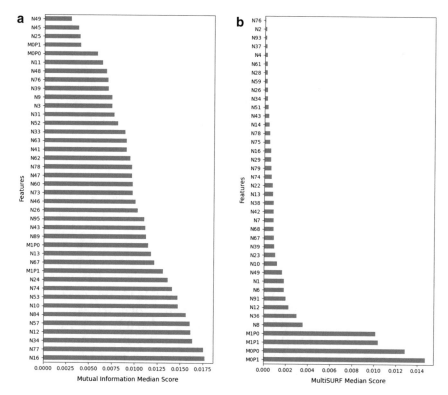

Fig. 2 Median feature importance results across 10-fold CV on simulated SNP data. The left barplot gives MI scores and the right barplot gives MultiSURF scores. Simulated predictive feature names start with an 'M' while non-predictive features start with an 'N'. Note how MultiSURF perfectly prioritizes the four features simulated with epistatic and heterogeneous patterns of the association while MI does not

STREAMLINE will first drop any features with a score $<= 0$ then alternate between each FS algorithm taking the unique, top-scoring features until the maximum is reached. Figure 2 presents median feature importance scores (over all CVs) for the two FS algorithms applied to the aforementioned example data. This highlights the potential value of collective feature selection, given that if we had only used MI, some of the 4 predictive features (M0P0, M0P1, M1P0, and M1P1) may have been removed based on how the maximum features to keep were set prior to modeling.

MultiSURF scales quadratically with the number of training instances; therefore, this notebook includes a user parameter to control the maximum number of randomly selected instances used by MultiSURF during scoring (set to 2000 by default to avoid very large run-times).

Of note, it has been previously demonstrated that when there are 10,000 features or more, MultSURF begins to fail to properly rank features involved in a pure interaction. However, Relief wrapper methods such as TuRF [34] have been proposed that can boost ranking success by iteratively filtering out the poorest scoring features over subsequent runs of the Relief-based algorithm. STREAMLINE includes the option to apply TuRF, which we recommend using in datasets with > 10,000 features.

FS takes place independently within each of the k training sets meaning that it is possible for each set to have a different subset of features available for model training. Notably, STREAMLINE exports the training and testing sets generated after CV partitioning, imputation, scaling, and feature selection so users can easily apply other algorithms and analyses to these same sets outside the pipeline.

2.1.3 Modeling

The heart of STREAMLINE is ML modeling itself. ML includes a large family of algorithmic methodologies that differ with respect to knowledge representation and inductive learning approaches as well as to the types of data and associations they are most effectively applied to and how interpretable their respective models will be. This first release of STREAMLINE includes a total of 15 scikit-learn compatible implementations of ML classification algorithms that can be individually utilized or left out. This set was selected based on their general popularity, ease of pipeline integration, and offer variety with respect to model representation and learning approaches. These include Naive Bayes (NB) [12], logistic regression (LR) [11], decision tree (DT) [8], random forest (RF) [13], gradient boosting trees (GB) [9], extreme gradient boosting (XGB) [2], light gradient boosting (LGB) [4], catboost (CGB) [1], support vector machines (SVM) [14], artificial neural networks (ANN) [7], k-nearest neighbors (KNN) [10], genetic programming (GP) symbolic classifier using 'gp-learn' [3], educational learning classifier system (eLCS) [5, 56], 'X' classifier system (XCS) [15], and an extended supervised tracking and classifying system (ExSTraCS) [6, 49, 55]. ExSTraCS is a RBML algorithm developed by our research group which combines evolutionary learning with a 'piece-wise' knowledge representation comprised of a set of IF:THEN rules. Unlike most ML methods, many RBML algorithms learn iteratively from instances in the training data and adapt a population of IF:THEN rules to cover unique parts of the problem space making them particularly well suited to capturing heterogeneous patterns of association [49]. RF, GB, XGB, LGB, and CGB are all decision tree-based algorithms and are currently among the most popular and successful ML algorithms within the ML research community. The implementation of GP we selected for inclusion is scikit-learn compatible, accessible, and simple to use. We included GP in STREAMLINE to highlight it as one of the few directly interpretable modeling algorithms. However the scikit-learn implementation utilized is basic, and may not reflect the potential performance of more sophisticated and modern GP algorithms.

Notably, many algorithms have other implementation options available which might be better tuned to specific problem domains or yield better performance in general. As such any ML comparison results from STREAMLINE, or any other AutoML tool that relies on specific implementations, should not be casually interpreted as a demonstration of the general superiority of one algorithm over another. By default, STREAMLINE will run all algorithms with the exception of eLCS and XCS which are still actively being developed and have known issues identified in early testing of this pipeline. This also highlights how STREAMLINE can be an effective development tool to test, debug, and evaluate new ML algorithm implementations.

STREAMLINE intentionally does not automate 'algorithm selection' as it seeks to present and compare ML algorithm modeling performance in as transparent and unbiased a manner as possible across a wide variety of evaluation metrics. This is intended to give users greater insight with respect to findings and greater flexibility with respect to how AutoML results can be utilized.

Hyperparameter Optimization: According to current best practices, the first step in ML modeling is to conduct a hyperparameter optimization sweep, a.k.a. 'tuning' [40]. Hyperparameters refer to the run parameters of a given ML algorithm that controls its functioning. Too often, ML algorithms are applied using their 'default' hyperparameter settings. This can lead to unfair ML algorithm comparisons, and a missed opportunity to obtain the best-performing model possible. Optimization effectively 'tries out' different hyperparameter settings on a subset of the training instances, ultimately selecting those yielding the best performance with which to train the final model. The first consideration is what hyperparameters to explore for each algorithm, as well as the range or selection of hyperparameter values to consider for each. As there is no clear consensus, we have surveyed a number of online sources, publications, and consulted colleagues in order to select the wide variety of hyperparameters and value ranges incorporated into STREAMLINE that are suited to binary classification. These hyperparameters and value options for each algorithm are detailed in the STREAMLINE Jupyter Notebook, as well as hard coded in the associated ModelJob.py script [16]. For example, STREAMLINE considers 8 hyperparameters for the optimization of the RF algorithm including 'the number of estimators', i.e. the number of trees in the 'forest', with potential values ranging from 10 to 1000. Users applying the Jupyter Notebook run mode can easily adjust the values considered in the hyperparameter optimization of each ML algorithm if desired; however, we have set these options to be as broadly applicable as possible.

The second consideration is what approach to take in conducting the hyperparameter sweep. Common approaches include a 'grid search' or a 'random search' [45] which either exhaustively or randomly consider hyperparameter value combinations. STREAMLINE adopts an 'automated' package called 'Optuna', which applies Bayesian optimization to try and 'intelligently' explore the specified hyperparameter space [17]. By default, a further 3-fold nested CV is conducted on the training dataset using a primary evaluation metric, where the user specifies a target number of optimization trials to complete and a maximum timeout to more equally limit the amount of run time each algorithm can spend on optimization. However, in order for STREAMLINE to ensure complete reproducibility, particularly when run in parallel, the user must specify 'None', for the Optuna timeout parameter, since variations in computing speed can lead to a different number of trials having been completed for each run. By default, all algorithms utilize hyperparameter optimization with the exception of Naive Bayes (which has no hyperparameters) as well as eLCS, XCS, and ExSTraCS which are more computationally expensive, in general,

Parallel Coordinate Plot

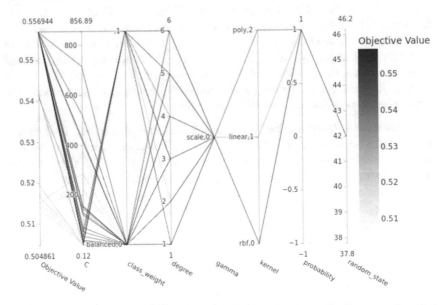

Fig. 3 Optuna-generated visualization of the SVM algorithm hyperparameter sweep conducted on one CV partition of the example simulated genomics data

and have relatively reliable default settings and reasonable guidelines to choose non-default hyperparameter settings as needed. This likely puts ExSTraCS at somewhat of a disadvantage compared to other ML algorithms in this pipeline.

STREAMLINE also provides users with the option to output hyperparameter sweep visualizations to illustrate how different settings combinations impact performance (see Fig. 3). Note how the sweep suggests that the 'kernel' hyperparameter (on the x-axis) performs poorly when set to 'linear', performs better when set to 'poly' and best when set to 'rbf', i.e. radial basis function, when applied to this data.

Train 'Optimized' Model: Once 'optimal' hyperparameters have been selected for each of the k training datasets and ML algorithms, a 'final' model is trained for each using the full training set and respective selected hyperparameters. The resulting models are stored as 'pickled' Python objects so that they can be easily utilized later.

Model Feature Importance Estimation: Next, this pipeline further examines feature importance, now from the perspective of each ML model. These estimates offer useful insights for high-level interpretation of models. Specifically, which features were most important for making accurate predictions. Some algorithm implementations including LR, DT, RF, GB, XGB, LGB, CGB, eLCS, XCS, and ExSTraCS have built-in strategies to report feature importance estimates after model training. In order to obtain feature importance estimates for the other 5 algorithms, this pipeline implements permutation feature importance in scikit-learn. This strategy randomly permutes one feature value at a time in the testing data to evaluate its impact on the target performance metric. STREAMLINE uniformly utilizes permutation feature importance for all algorithms by default in order to provide a consistent metric for comparison. However, users can choose to use built-in algorithm strategies whenever available and permutation feature importance for the other 5 algorithms as an alternative.

Any feature that was removed by feature selection prior to modeling is given a model feature importance estimate of zero by default in the respective CV partition.

Evaluate Performance: The last element in ML modeling is to evaluate model performance. It is essential to select appropriate evaluation metrics to fit the characteristics and goals of the given analysis. STREAMLINE ensures that a comprehensive selection of binary classification metrics and visualizations are employed that offer a holistic perspective of model performance. Earlier elements of this pipeline call for a target evaluation metric, i.e. hyperparameter optimization and model feature importance estimation. STREAMLINE uses *balanced accuracy* by default since this metric equally emphasizes accurate predictions within both classes. In addition to balanced accuracy, this notebook calculates and reports the following 15 evaluation metrics: true positive count, true negative count, false positive count, false negative count, standard accuracy, F1-Score, sensitivity/recall, specificity, precision, receiver operating characteristic (ROC), area under the curve (AUC), precision-recall curve (PRC) AUC, PRC average precision score (APS), negative predictive value, positive likelihood ratio, and negative likelihood ratio. STREAMLINE saves model performance metrics, feature importance estimates, and can output prediction probabilities (post-hoc) on testing or replication data as .csv files and 'pickled' objects (to facilitate future use) for each algorithm and CV partition.

2.1.4 Post-analysis

STREAMLINE post-analysis calculates mean and median performance results (over all CV partition models), generates figures, conducts non-parametric statistical comparisons, generates a PDF summary report of key findings, and allows easy application of all trained models to available replication data or other future data to further evaluate model generalization capability.

Performance Summary: STREAMLINE calculates CV means, medians, and standard deviations for all evaluation metrics and model feature importance scores.

Figure Generation: STREAMLINE automatically generates a wide variety of figures capturing performance and promoting model interpretability. For each algorithm, STREAMLINE pipeline outputs an ROC and PRC plot summarizing the performance of each of the k trained models alongside the mean ROC or PRC, respectively. PRC plots are preferable to ROC plots when class imbalance is more extreme or if the practitioner is more concerned with making positive class predictions. This pipeline automatically sets the 'no-skill' lines of PRC plots based on the

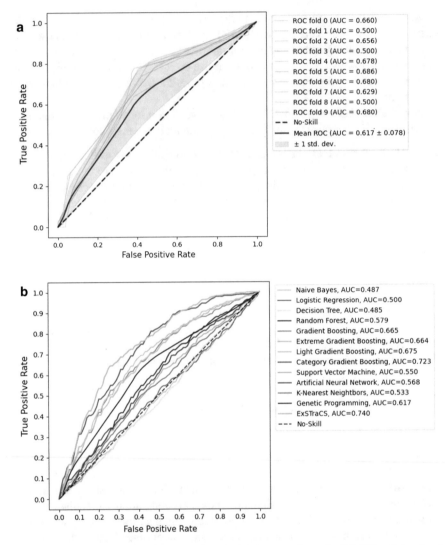

Fig. 4 ROC plots for the example simulated genomics dataset: **a** for each of 10 CV models trained with the GP algorithm, and **b** for average performance of the 13 ML algorithms that STREAMLINE runs by default. Note that our rule-based ML algorithm, ExSTraCS, performs at least as well as all other algorithms in this example despite not having the benefit of hyperparameter optimization. This illustrates how STREAMLINE may be applied to verify the efficacy of new ML algorithms

class ratios in the dataset. Additionally, ROC and PRC plots comparing average CV performance across all algorithms are generated. Figure 4 presents ROC plots and Fig. 5 presents PRC plots illustrating the comparison of (A) individual CV runs of the GP algorithm and (B) averaged performance of all algorithms.

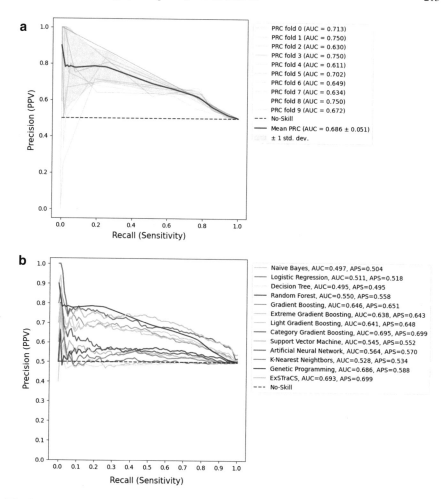

Fig. 5 PRC plots for the example simulated genomics dataset: **a** for each of 10 CV models trained with the GP algorithm, and **b** for the average performance of the 13 ML algorithms that STREAM-LINE runs by default

STREAMLINE also automatically generates (1) boxplots for each evaluation metric comparing performance across all ML algorithms, (2) boxplots of top, ranked model feature importance estimates for each ML algorithm, and (3) histograms of feature importance score distribution for each ML algorithm (not shown here). STREAMLINE also generates proposed 'composite feature importance bar plots' (CFIBP) that illustrate and summarize model feature importance consistency across all ML algorithms. The focal CFIBP generated normalizes feature importance scores within each algorithm between 0 and 1 and then weights the normalized scores by the median balanced accuracy or median ROC-AUC of the respective algorithm so that those which do not perform as well have less impact on the visualization. Figure 6

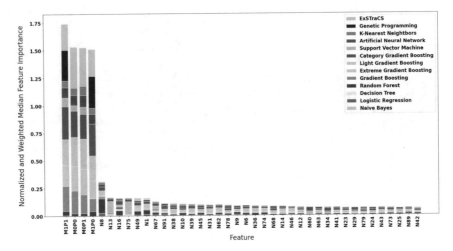

Fig. 6 Composite feature importance bar plot (CFIBP) for the example simulated genomics dataset (top 40 features). Note how top-performing algorithms correctly identify M0P0, M0P1, M1P0, and M1P1 as most important, but (in this data scenario) the RF algorithm incorrectly prioritizes N8, having performed less well in comparison

gives the CFIBP generated for the example genomics dataset, illustrating the consensus of top-performing algorithms in identifying the true predictive features.

If STREAMLINE is applied to multiple datasets in an 'experiment', it will create boxplots for each metric comparing average results across all ML algorithms including lines identifying how the performance of each ML algorithm changes. It will create similar boxplots for each ML algorithm and metric combination comparing CV runs from one dataset to the next. Examples of these dataset comparison figures are included in Sect. 3.

Included in the STREAMLINE GitHub repository [16] is a folder of 'Useful Notebooks', i.e. Jupyter Notebooks designed to facilitate an array of additional tasks following the main pipeline analysis including (1) exporting prediction probabilities for training, testing, or replication data, (2) making custom modifications to key figures, (3) exploring and evaluating different decision thresholds beyond the standard 0.5, (4) generating a ranked interactive model feature importance heat-map (see Fig. 7), and (5) generating direct visualizations of the most interpretable classification algorithms, i.e. DT and GP (see Figs. 8 and 9).

Statistical Comparisons: STREAMLINE is set up to automatically conduct non-parametric statistical comparisons (1) between ML algorithms using CV partitions, and (2) (if more than one dataset is being analyzed) between datasets for each individual ML algorithm as well among the best-performing ML algorithms for every evaluation metric. In either scenario, STREAMLINE first applies Kruskal-Wallis one-way analysis of variance to determine if any ML algorithm or datasets yielded significantly different performance than the others for each evaluation metric. For any metric where a significant difference was observed, STREAMLINE follows up

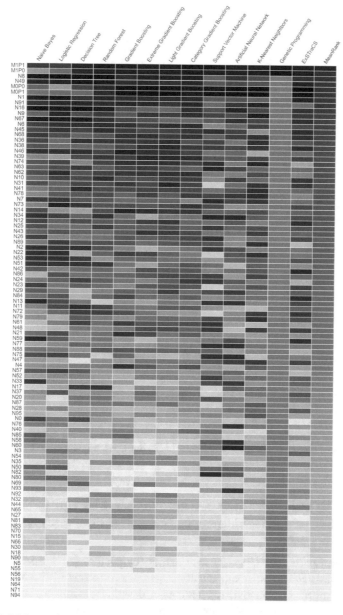

Fig. 7 Model feature importance heat-map with average CV ranking across all ML algorithms. Dark blue denotes higher importance and bright yellow denotes low importance. This interactive heat-map can be opened as a web-page generated by one of the STREAMLINE 'Useful Notebooks'. Placing the cursor over any cell (when opened as an 'html') will reveal the algorithm, feature name, and rank of that feature for the given algorithm

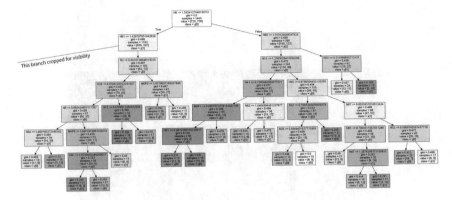

Fig. 8 Direct visualization of the DT model with the highest CV testing ROC-AUC for the example simulated genomics dataset. As seen in Fig. 4, DT performs quite poorly on this dataset, and this overly complicated DT tree is struggling to detect any associations in the presence of heterogeneous feature interactions

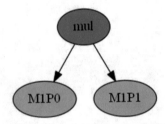

Fig. 9 Direct visualization of the GP model with the highest CV testing ROC-AUC for the example simulated genomics dataset. As seen in Fig. 4, GP performs only moderately well on this dataset compared to some other algorithms. This simple and clearly interpretable GP tree is picking up on one of the two heterogeneous 2-way feature interactions simulated in the dataset suggesting that this GP implementation can tackle 2-way feature interactions, but struggles to model heterogeneous relationships

with pairwise Mann-Whitney U-tests and Wilcoxon Rank tests for that metric to identify which pairs of algorithms or datasets yielded significantly better or worse performance.

PDF Summary Report: Given the wide range of output files generated and organized by STREAMLINE, it also automatically generates a formatted PDF summary report laying out all STREAMLINE run settings, a descriptive summary of key dataset characteristics, evaluation and feature importance plots, and metric results for each dataset, as well as Kruskal-Wallis significance results comparing top algorithms for each metric/dataset combination, and runtime of all elements and specific ML algorithms. A report is generated for each STREAMLINE 'experiment' as well as for any application of trained models to any additional replication datasets.

Application to Replication Data: Once STREAMLINE has run an 'experiment', all trained models can be evaluated on additional hold-out replication data to help verify model generalizability and globally pick a 'best' model. This 'apply-model' phase of STREAMLINE automatically loads pickled models, as well as information on how to impute and scale the data in line with the original training data. This can be run on any number of additional replication datasets. If a user wishes to select a final model to put into practice as a predictive model in real-world applications such as clinical decision support, we recommend using the results of applying all models to the replication data in order to choose it. This (1) ensures that all models are evaluated and compared using the same hold-out data, (2) reduces the opportunity for sample bias, and (3) provides further rigor in the assessment of model generalizability. Differently, outside of STREAMLINE, users could create an ensemble model (as their final model) that combines the predictions of all CV models for a single algorithm or that combines all models (algorithms and CV partitions) trained by the pipeline. Automating ensemble prediction as STREAMLINE option will be included in a future release.

2.1.5   Implementation and Run Modes

STREAMLINE was implemented using Python 3 and a variety of well-established Python packages including scikit-learn, pandas, scipy, numpy, optuna, and pickle. STREAMLINE has 4 available run modes designed to suit users with different levels of computing experience and analysis needs. We review these modes below, but direct users to the STREAMLINE README for specific instructions for installation and use [16].

Mode 1 (Google Colab Notebook): This is the easiest way to run STREAMLINE, requiring no coding or computing environment experience, using a Google Colab Notebook. This runs the pipeline serially using Google Colab and Google Cloud for computing. This is currently the slowest and most computationally limited option. It's ideal for those new to ML looking to learn or run simpler, smaller-scale analyses.

Mode 2 (Jupyter Notebook): For those with a little Python experience, and are familiar with installing Anaconda, and the commands to install other Python packages. This option offers users easy access to modify STREAMLINE run parameters and ML algorithm hyperparameter options. This option also runs serially but gives users direct access to results and figures within the notebook as they are generated.

Mode 3 (Local Command Line): For those with command line experience this run mode is likely preferable. The 4 stages of STREAMLINE, described above, are set up to run (in sequence) as 11 possible command line phases optimized (in particular) for Mode 4 parallelization. This is detailed in [16].

Mode 4 (Computing Cluster Parallelized Command Line): For those with access to an LSF compatible Linux computing cluster, STREAMLINE is currently set up to efficiently parallelize its 4 stages as 11 compute phases. Programming savvy users should be able to easily update the 'Main' Python scripts to parallelize STREAM-LINE analyses using other distributed computing environments such as Amazon Web Services, Microsoft Azure, Google Cloud, or other local compute clusters.

2.2 Evaluation Datasets

2.2.1 HCC UCI Benchmark Data

In this paper, we first evaluated STREAMLINE performance on the hepatocellu-
lar carcinoma (HCC) survival benchmark dataset from the UCI repository [21]. We
chose the HCC dataset as the demonstration dataset, included with the STREAM-
LINE software, as it's small, i.e. 165 instances, and 49 features, thus quick to analyze, and because it exemplifies many of the standard data considerations/challenges
including having (1) a mix of feature types (i.e. categorical and numerical), (2) about
10% missing values, and (3) class imbalance with 63 patients who were deceased and
102 who survived. For the purposes of demonstration, we created a second dataset
from this HCC data that removed two common covariates (i.e. age and sex) to show
how STREAMLINE could be applied to multiple datasets at once in order to easily
compare dataset performance in this context. We repeated this analysis twice in parallel using a fixed number of Optuna trials and no time limit to confirm the replicability
of STREAMLINE. To test the 'apply-model' functionality of STREAMLINE, we
used the entire HCC dataset as a stand-in replication dataset, since a true replication
dataset wasn't available and training instances were limited.

2.2.2 GAMETES Simulated Datasets

Next, we evaluated 6 other genomics datasets simulated with GAMETES [47],
each with a different underlying modeled pattern of association to illustrate how
STREAMLINE can be utilized to evaluate the strengths and weaknesses of different
ML algorithms in different contexts. Each dataset simulated 100 single nucleotide
polymorphisms (SNPs) as features and included 1600 instances using a minor allele
frequency of 0.2 for relevant features and the 'easiest' model architecture generated
for each configuration using the GAMETES simulation approach [46]. All irrelevant
features were randomly simulated with a minor allele frequency between 0.05 and
0.5. We detail dataset differences here (A) Univariate association between a single
feature and outcome with 1 relevant and 99 irrelevant features and a heritability of
0.4, (B) Additive combination of 4 univariate associations with 4 relevant and 96
irrelevant features and heritability of 0.4 for each relevant feature, (C) Heterogeneous combination of 4 univariate associations, i.e. each univariate association is
only predictive in a respective quarter of instances (with all else the same as dataset
B), (D) Pure 2-way feature interaction with 2 relevant and 98 irrelevant features
and heritability of 0.4, (E) Heterogeneous combination of 2 independent pure 2-way
feature interactions with 4 relevant and 96 irrelevant features and heritability of 0.4
for each 2-way interaction (used in above sections for figure examples), and (F) Pure
3-way feature interaction with 3 relevant and 97 irrelevant features and heritability
of 0.2 (i.e. the most difficult dataset).

When STREAMLINE conducts modeling in the datasets discussed so far, ExS-TraCS, which was trained without a hyperparameter sweep, is set to run with hyperparameters ($nu = 1$, rule population size $=2K$, and training iterations $= 200K$), where the nu setting represents the emphasis on discovering and keeping rules with maximum accuracy. Previous work suggests that $nu = 1$ is more effective in noisy data, while a nu of 5 or 10 is more effective in clean data (i.e. no noise).

2.2.3 *x*-bit MUX Benchmark Datasets

Lastly, we applied STREAMLINE to 6 different *x*-bit multiplexer (MUX) binary classification benchmark datasets (i.e. 6, 11, 20, 37, 70, and 135-bit) often utilized to evaluate ML algorithms such as RBML [49]. Like GAMETES dataset 'E', these MUX datasets involve both feature interactions and heterogeneous associations with outcome, but differently involve binary features and a clean association with outcome. The '*x*'-bit value denotes the number of relevant features underlying the association, and increasing values from 6 to 135 dramatically scale up the complexity of the underlying pattern of association. Specifically, solving the 6-bit problem involves modeling 4 independent 3-way interactions, while the 135-bit problem involves modeling 128 independent 8-way interactions. In these datasets, all features are relevant to solving the problem, where a subset of features serves as 'address bits' which point to one of the remaining 'register bits'. The value of that corresponding register bit indicates the correct outcome (i.e. 0 or 1).

In previous work, the ExSTraCS algorithm was the first ML demonstrated to directly model the 135-bit problem successfully [49], in a dataset including 40K training instances, with hyperparameters ($nu = 10$, rule population size $=10K$, and training iterations $= 1.5$ million). In this study, 6 to 135-bit datasets were generated with 500, 1000, 2000, 5000, 10000, and 20000 instances, respectively, with 90% of each used by STREAMLINE for training (due to 10-fold CV). In previous work, it was demonstrated that larger numbers of instances are required to solve or nearly solve increasingly complex MUX problems. In contrast with modeling on the noisy datasets, for the MUX datasets, ExSTraCS was assigned the following hyperparameter settings: $nu = 10$, rule population size $=5K$, and training iterations $= 500K$.

All analyses were conducted in parallel on an LSF compatible, Linux computing cluster with Anaconda3-2022.05-Linux-x86_64, and all additional Python packages installed as described in the Beta 0.2.4 release of STREAMLINE [16].

3 Results and Discussion

In this section, we summarize high-level findings of STREAMLINE experiments on the datasets described in the previous section. As should be apparent from the description of STREAMLINE above, there are many results and figures that could be presented; therefore, instead of including them here, we have made the STREAM-

LINE PDF report summaries for each 'experiment' available at https://github.com/UrbsLab/STREAMLINE/tree/main/Experiments. Documentation of all pipeline settings for respective experiments is included on the first page of each report.

3.1 HCC UCI Benchmark Data Results

This analysis was designed to illustrate the overall functionality of the entire pipeline on a small, simple dataset. We repeated this STREAMLINE experiment on the same set of two HCC datasets with the same pipeline configurations and random seed to confirm complete reproducibility (see aforementioned PDF experiment summaries on GitHub). Focusing on the original HCC dataset, Fig. 10 compares ROC and PRC ML algorithm performance. While SVM yielded the highest average ROC-AUC and CGB yielded the highest average, PRC-AUC Kruskal-Wallis testing indicated these differences were not significant, likely in part due to the small sample size available in this analysis. A review of mean and median metric performance across all algorithms highlights how there is no single clear algorithmic 'winner' across all metrics.

Figure 11 gives model feature importance estimates across all ML algorithms, indicating the feature, Ferritin (ng/ML), to be most consistently informative. In contrast with pre-modeling feature importance estimates, this feature had previously only been ranked 6th by MI and 11th by MultiSURF.

Lastly, Fig. 12 compares the performance of the ML algorithm between the full HCC dataset and the same data with the covariate features removed. This pairing of boxplots illustrates how investigators can use STREAMLINE output to look more closely at specific performance differences across and within ML algorithms and datasets.

3.2 GAMETES Simulated Datasets Results

This analysis was designed to highlight the basic advantages and disadvantages of ML algorithms in different GAMETES-simulated data scenarios. Figure 13 compares average ROC-AUC performance across all 6 GAMETES simulations. We observe fairly consistent performance across algorithms for the first two 'easiest' datasets and start to observe increasingly dramatic performance differences as heterogeneity and feature interactions are introduced.

We focus on mean ROC-AUC in the following performance discussion. For dataset (A), i.e. univariate, all algorithms performed similarly well, with KNN performing the least well. For dataset (B), i.e. additive univariate, all algorithms performed similarly well, with NB performing the least well. For dataset (C), i.e. heterogeneous univariate, CGB, XGB, RF, GB, and ExSTraCS performed best, while NB, KNN, and GP performed the least well. For dataset (D), i.e. pure 2-way epistasis, we observe the first dramatic performance differences with ExSTraCS, CGB, LGB, XGB, and

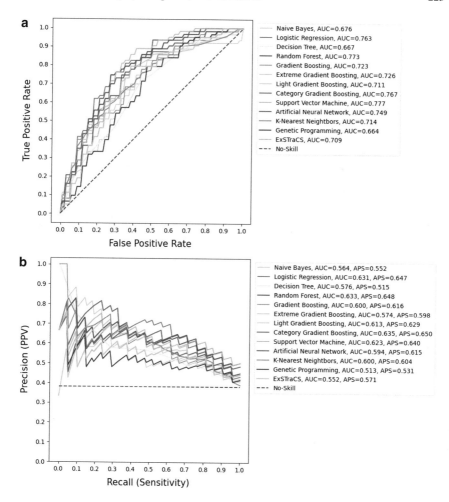

Fig. 10 Full HCC dataset analysis: **a** ROC plot and **b** PRC plot, each illustrating averages over all CV models

GB performing similarly at the top, while NB, LR, and DT entirely failed to detect the 2-feature interaction. For dataset (E), i.e. heterogeneous pure 2-way epistasis, ExSTraCS and CGB stood out as top performers, while again NB, LR, and DT failed. Lastly, for dataset (F), i.e. pure 3-way epistasis, all algorithms struggled with this noisy complex association, but ExSTraCS performed slightly but significantly better across a number of metrics. Across these simulated scenarios, ExSTraCS performed consistently well and often best of all ML algorithms included highlighting the competitive efficacy of evolutionary rule-based machine learning. However, as would be expected, no algorithm stands out as being ideal across all data scenarios, or evaluation metrics.

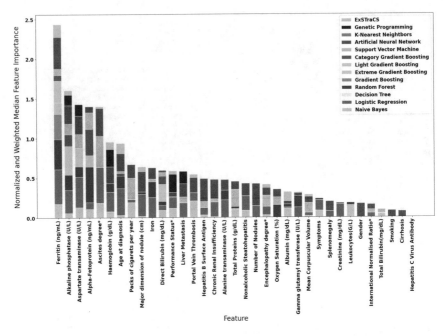

Fig. 11 CFIBP of the full HCC dataset (top 40 features)

For all datasets, model feature importance CFIBPs correctly differentiated simulated relevant from irrelevant features.

3.3 x-*bit MUX Benchmark Datasets Results*

This analysis was designed to examine how different ML algorithms performed when modeling data with an increasing underlying problem complexity as well as to highlight how other well-known ML algorithms compare to ExSTraCS (previously shown to perform well on these benchmark problems). Figure 14 compares performance across all 6 MUX datasets. The simplest, 6-bit MUX is solved perfectly (i.e. ROC-AUC = 1, averaged over CV partitions) by all algorithms except NB, LR, and GP, although GP achieved an ROC-AUC of 0.979. These three algorithms failed to solve all subsequent MUX datasets. For the 11-bit MUX, only the ensemble tree-based ML algorithms (i.e. RF, GB, XGB, LGB, and CGB) and ExSTraCS performed perfectly, and other algorithms yielded deteriorating performance. GP in particular yielded a dramatic performance drop between the 6 and 11-bit MUX. For the 20-bit MUX, only CGB and ExSTraCS yield perfect performance, although GB, XGB, and LGB yielded near-perfect performance, and RF achieves an ROC-AUC of 0.987. Figure 15 presents the CFIBP for the 20-bit MUX, highlighting the expected increased feature

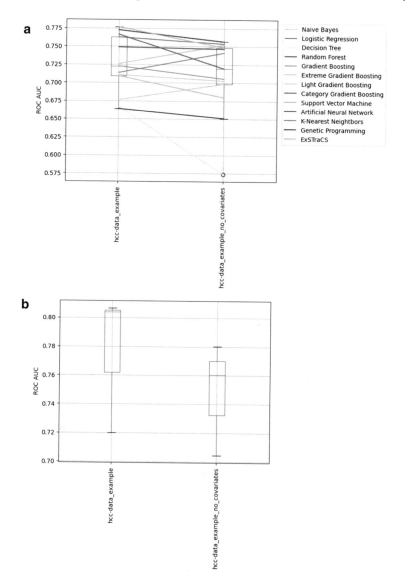

Fig. 12 Comparing performance on HCC data with and without covariates. **a** Comparing average ML ROC-AUC between all algorithms. The average performance shift of each algorithm is indicated by the respective line connecting the boxplots. **b** Comparing individual CV runs of SVM which yielded the best ROC-AUC on the full HCC dataset. A clear but non-significant reduction in performance was observed when covariates were removed

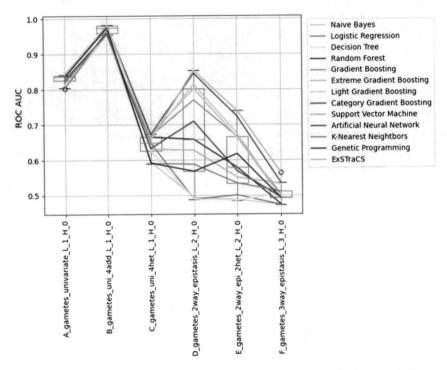

Fig. 13 Comparing average ML ROC-AUC performance across GAMETES datasets (**A–F**)

importance estimates for address bit in contrast with register bit features. For the 37-bit MUX, only ExSTraCS yielded perfect performance, although GB, XGB, and LGB yielded ROC-AUCs over 0.9. For the 70-bit MUX, ExSTraCS performed dramatically better than all other algorithms with an ROC-AUC of 0.992. Lastly, for the 135-bit MUX, all algorithms failed to perform well, with LGB performing best with an ROC-AUC of 0.559. This failure was largely expected as only 18K instances were available for model training, when previous work required 40K instances for ExSTraCS to closely solve the 135-bit MUX where a staggering $4.36e40$ unique binary instances make up the problem space. In this particular benchmark analysis, ExSTraCS stands out clearly as the best-performing algorithm. Overall this analysis highlights the capability of RBML algorithms such as ExSTraCS to perform competitively and sometimes better than other well-established ML modeling approaches. Additionally, it provides another example of how STREAMLINE facilitates the comparison of ML algorithm performance as a structured, rigorous analysis.

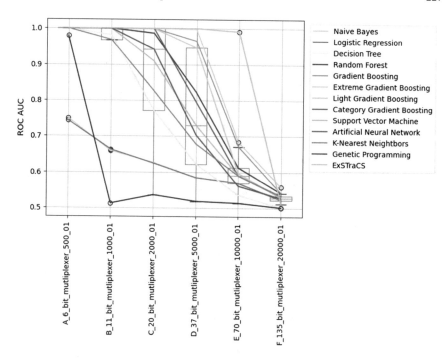

Fig. 14 Comparing average ML ROC-AUC performance across 6, 11, 20, 37, 70, and 135-bit multiplexer benchmark datasets

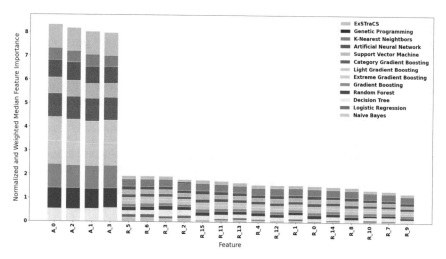

Fig. 15 CFIBP of 20-bit MUX. This dataset includes 4 address bits 'A' and 16 register bits 'R'. In MUX problems, address bits are relevant in predicting all instances, but register bits are only relevant in the subset of instances where the address bits point to them. This is why address bits have a much larger feature importance than register bits in this plot. In the 20-bit MUX, 16 separate 5-way interactions, each involving 4 addresses and 1 register bit, must be found to solve the problem

4 Conclusions

In this paper, we introduced STREAMLINE, a simple transparent end-to-end AutoML pipeline. The initial application of STREAMLINE to a variety of benchmark datasets illustrates how this tool can be used beyond a standard data analysis to facilitate comparison of (1) new to established ML modeling algorithms, (2) ML performance across different target datasets, and (3) other autoML tools to rigorous benchmark. In this first release, we sought to include a mix of well-known ML algorithms and a handful of newer or less well-known algorithms with unique advantages (i.e. genetic programming and rule-based machine learning). These findings in applying STREAMLINE support the importance of utilizing a variety of ML modeling algorithms, each with different strengths and weaknesses.

Future work will target many expansions and new applications for STREAMLINE including (1) extending the framework to accommodate multi-class and regression analysis, (2) adding new algorithms, visualizations, and automated elements to the pipeline including one-hot encoding for categorical variables and Shapley values for model interpretation, (3) providing support to run STREAMLINE in different parallel computing environments, (4) adding environment controls such as 'Docker' to further facilitate reproducibility, (5) applying it to comprehensively evaluate new feature selection and ML modeling methodologies, (6) applying it as a positive control benchmark to evaluate potential performance advantages using other AutoML or artificial-intelligence-driven ML tools, (7) identifying a subset of ML algorithms that can be reliably applied in different scenarios to enhance understanding of a given dataset, (8) leveraging this platform as an educational and algorithm development tool by facilitating overall analysis comprehension and transparency, and (9) utilizing the STREAMLINE framework to encourage a broader adoption of promising, less well-known strategies such as GP or rule-based ML in the future.

Acknowledgements The study was supported by the following NIH grants: R01s LM010098 and AG066833. STREAMLINE development benefited from multiple biomedical research collaborators at the University of Pennsylvania, Fox Chase Cancer Center, Cedars Sinai Medical Center, and the University of Kansas Medical Center. Special thanks to Patryk Orzechowski, Trang Le, Sy Hwang, Richard Zhang, Wilson Zhang, and Pedro Ribeiro for their code contributions and feedback. We also thank the following collaborators for their feedback on the application of the pipeline during development: Shannon Lynch, Rachael Stolzenberg-Solomon, Ulysses Magalang, Allan Pack, Brendan Keenan, Danielle Mowery, Jason Moore, and Diego Mazzotti.

References

1. Catboost. https://catboost.ai/en/docs/. Accessed 28 May 2022
2. Extreme gradient boosting. https://xgboost.readthedocs.io/en/stable/. Accessed 28 May 2022
3. gp-learn github respository. https://github.com/trevorstephens/gplearn. Accessed 28 May 2022
4. Light gradient boosting. https://lightgbm.readthedocs.io/en/latest/. Accessed 28 May 2022
5. scikit-elcs github respository. https://github.com/UrbsLab/scikit-eLCS. Accessed 28 May 2022

6. scikit-exstracs github respository. https://github.com/UrbsLab/scikit-ExSTraCS. Accessed 28 May 2022

7. scikit-learn ann. https://scikit-learn.org/stable/modules/generated/sklearn.neural_network. MLPClassifier.html. Accessed 28 May 2022

8. scikit-learn decision tree. https://scikit-learn.org/stable/modules/generated/sklearn.tree. DecisionTreeClassifier.html. Accessed 28 May 2022

9. scikit-learn gradient boosting trees. https://scikit-learn.org/stable/modules/generated/sklearn. ensemble.GradientBoostingClassifier.html. Accessed 28 May 2022

10. scikit-learn knn. https://scikit-learn.org/stable/modules/generated/sklearn.neighbors. KNeighborsClassifier.html. Accessed 28 May 2022

11. scikit-learn logistic regression. https://scikit-learn.org/stable/modules/generated/sklearn. linear_model.LogisticRegression.html. Accessed 28 May 2022

12. scikit-learn naive bayes. https://scikit-learn.org/stable/modules/generated/sklearn.naive_ bayes.GaussianNB.html. Accessed 28 May 2022

13. scikit-learn random forest. https://scikit-learn.org/stable/modules/generated/sklearn. ensemble.RandomForestClassifier.html. Accessed 28 May 2022

14. scikit-learn svm. https://scikit-learn.org/stable/modules/generated/sklearn.svm.SVC.html. Accessed 28 May 2022

15. scikit-xcs github respository. https://github.com/UrbsLab/scikit-XCS. Accessed 28 May 2022

16. Streamline github repository. https://github.com/UrbsLab/STREAMLINE. Accessed 28 May 2022

17. Akiba, T., Sano, S., Yanase, T., Ohta, T., Koyama, M.: Optuna: a next-generation hyperparameter optimization framework. In: Proceedings of the 25th ACM SIGKDD International Conference on Knowledge Discovery and Data Mining, pp. 2623–2631 (2019)

18. Buuren, S., Groothuis-Oudshoorn, K.: Mice: multivariate imputation by chained equations in r. J. Stat. Softw. **45**(3) (2011)

19. Chauhan, K., Jani, S., Thakkar, D., Dave, R., Bhatia, J., Tanwar, S., Obaidat, M.S.: Automated machine learning: The new wave of machine learning. In 2020 2nd International Conference on Innovative Mechanisms for Industry Applications (ICIMIA), pp. 205–212. IEEE (2020)

20. Diao, J.A., Kohane, I.S., Manrai, A.K.: Biomedical informatics and machine learning for clinical genomics. Human Molecul. Genet. **27**(R1), R29–R34 (2018)

21. Dua, D., Graff, C.: UCI machine learning repository (2017)

22. Elsebakhi, E., Lee, F., Schendel, E., Haque, A., Kathireason, N., Pathare, T., Syed, N., Al-Ali, R.: Large-scale machine learning based on functional networks for biomedical big data with high performance computing platforms. J. Comput. Sci. **11**, 69–81 (2015)

23. Fabris, F., Freitas, A.A.: Analysing the overfit of the auto-sklearn automated machine learning tool. In: International Conference on Machine Learning, Optimization, and Data Science, pp. 508–520. Springer (2019)

24. Garreta, R., Moncecchi, G., Hauck, T., Hackeling, G.: Scikit-Learn: Machine Learning Simplified: Implement Scikit-Learn into Every Step of the Data Science Pipeline. Packt Publishing Ltd, 2017

25. Greener, J.G., Kandathil, S.M., Moffat, L., Jones, D.T.: A guide to machine learning for biologists. Nat. Rev. Molecul. Cell Biol. **23**(1), 40–55 (2022)

26. Heil, B.J., Hoffman, M.M., Markowetz, F., Lee, S.-I., Greene, C.S., Hicks, S.C.: Reproducibility standards for machine learning in the life sciences. Nat. Methods **18**(10), 1132–1135 (2021)

27. Hutter, F., Kotthoff, L., Vanschoren, J.: Automated Machine Learning: Methods, Systems. Challenges, Springer Nature (2019)

28. Krstajic, D., Buturovic, L.J., Leahy, D.E., Thomas, S.: Cross-validation pitfalls when selecting and assessing regression and classification models. J. Cheminformat. **6**(1), 1–15 (2014)

29. Kusiak, A.: Feature transformation methods in data mining. IEEE Trans. Electron. Packag. Manufact. **24**(3), 214–221 (2001)

30. La Cava, W., Williams, H., Fu, W., Vitale, S., Srivatsan, D., Moore, J.H.: Evaluating recommender systems for ai-driven biomedical informatics. Bioinformatics **37**(2), 250–256 (2021)

31. Linden, A., Yarnold, P.R.: Using machine learning to assess covariate balance in matching studies. J. Eval. Clin. Pract. **22**(6), 848–854 (2016)
32. Luo, J., Wu, M., Gopukumar, D., Zhao, Y.: Big data application in biomedical research and health care: a literature review. Biomed. Inf. Insights **8**, BII–S31559 (2016)
33. Luo, W., Phung, D., Tran, T., Gupta, S., Rana, S., Karmakar, C., Shilton, A., Yearwood, J., Dimitrova, N., Ho, T.B., et al.: Guidelines for developing and reporting machine learning predictive models in biomedical research: a multidisciplinary view. J. Med. Internet Res. **18**(12), e323 (2016)
34. Moore, J.H., White, B.C.: Tuning relieff for genome-wide genetic analysis. In: European Conference on Evolutionary Computation, Machine Learning and Data Mining in Bioinformatics, pp. 166–175. Springer (2007)
35. Olson, R.S., Moore, J.H.: Tpot: a tree-based pipeline optimization tool for automating machine learning. In: Automated Machine Learning, pp. 151–160. Springer (2019)
36. Pedregosa, F., Varoquaux, G., Gramfort, A., Michel, V., Thirion, B., Grisel, O., Blondel, M., Prettenhofer, P., Weiss, R., Dubourg, V., et al.: Scikit-learn: machine learning in python. J. Mach. Learn. Res. **12**, 2825–2830 (2011)
37. Peng, H., Long, F., Ding, C.: Feature selection based on mutual information: criteria of max-dependency, max-relevance, and min-redundancy. IEEE Trans. Pattern Anal. Mach. Intell. **27**, 1226–1238 (2005)
38. Rauschert, S., Raubenheimer, K., Melton, P., Huang, R.: Machine learning and clinical epigenetics: a review of challenges for diagnosis and classification. Clin. Epigenet. **12**, 1–11 (2020)
39. Riley, P.: Three pitfalls to avoid in machine learning (2019)
40. Schratz, P., Muenchow, J., Iturritxa, E., Richter, J., Brenning, A.: Hyperparameter tuning and performance assessment of statistical and machine-learning algorithms using spatial data. Ecol. Model. **406**, 109–120 (2019)
41. Smialowski, P., Frishman, D., Kramer, S.: Pitfalls of supervised feature selection. Bioinformatics **26**(3), 440–443 (2010)
42. Thornton-Wells, T.A., Moore, J.H., Haines, J.L.: Genetics, statistics and human disease: analytical retooling for complexity. TRENDS Genet. **20**(12), 640–647 (2004)
43. Truong, A., Walters, A., Goodsitt, J., Hines, K., Bruss, C.B., Farivar, R.: Towards automated machine learning: evaluation and comparison of automl approaches and tools. In: 2019 IEEE 31st International Conference on Tools with Artificial Intelligence (ICTAI), pp. 1471–1479. IEEE (2019)
44. Uçar, M.K., Nour, M., Sindi, H., Polat, K.: The effect of training and testing process on machine learning in biomedical datasets. Math. Probl, Eng (2020)
45. Uppu, S., Krishna, A.: Tuning hyperparameters for gene interaction models in genome-wide association studies. In: International Conference on Neural Information Processing, pp. 791–801. Springer (2017)
46. Urbanowicz, R.J., Kiralis, J., Fisher, J.M., Moore, J.H.: Predicting the difficulty of pure, strict, epistatic models: metrics for simulated model selection. BioData Mining **5**(1), 1–13 (2012)
47. Urbanowicz, R.J., Kiralis, J., Sinnott-Armstrong, N.A., Heberling, T., Fisher, J.M., Moore, J.H.: Gametes: a fast, direct algorithm for generating pure, strict, epistatic models with random architectures. BioData Mining **5**(1), 1–14 (2012)
48. Urbanowicz, R.J., Meeker, M., La Cava, W., Olson, R.S., Moore, J.H.: Relief-based feature selection: introduction and review. J. Biomed. Inf. **85**, 189–203 (2018)
49. Urbanowicz, R.J., Moore, J.H.: Exstracs 2.0: description and evaluation of a scalable learning classifier system. Evolut. Intell. **8**(2–3), 89–116 (2015)
50. Urbanowicz, R.J., Olson, R.S., Schmitt, P., Meeker, M., Moore, J.H.: Benchmarking relief-based feature selection methods for bioinformatics data mining. J. Biomed. Inf. **85**, 168–188 (2018)
51. Verma, S.S., Lucas, A., Zhang, X., Veturi, Y., Dudek, S., Li, B., Li, R., Urbanowicz, R., Moore, J.H., Kim, D., et al.: Collective feature selection to identify crucial epistatic variants. BioData Mining **11**(1), 5 (2018)

52. Vieira, S., Garcia-Dias, R., Pinaya, W.H.L.: A step-by-step tutorial on how to build a machine learning model. In: Machine Learning, pp. 343–370. Elsevier (2020)
53. Waring, J., Lindvall, C., Umeton, R.: Automated machine learning: review of the state-of-the-art and opportunities for healthcare. Artif. Intell. Med. **104**, 101822 (2020)
54. White, I.R., Daniel, R., Royston, P.: Avoiding bias due to perfect prediction in multiple imputation of incomplete categorical variables. Comput. Stat. Data Anal. **54**(10), 2267–2275 (2010)
55. Zhang, R., Stolzenberg-Solomon, R., Lynch, S.M., Urbanowicz, R.J.: Lcs-dive: an automated rule-based machine learning visualization pipeline for characterizing complex associations in classification (2021). arXiv preprint arXiv:2104.12844
56. Zhang, R.F., Urbanowicz, R.J.: A scikit-learn compatible learning classifier system. In: Proceedings of the 2020 Genetic and Evolutionary Computation Conference Companion, pp. 1816–1823 (2020)

Evolving Complexity is Hard

Alden H. Wright and Cheyenne L. Laue

Abstract Understanding the evolution of complexity is an important topic in a wide variety of academic fields. Implications of better understanding complexity include increased knowledge of major evolutionary transitions and the properties of living and technological systems. Genotype-phenotype (G-P) maps are fundamental to evolution, and biologically-oriented G-P maps have been shown to have interesting and often-universal properties that enable evolution by following phenotype-preserving walks in genotype space. Here we use a digital logic gate circuit G-P map where genotypes are represented by circuits and phenotypes by the functions that the circuits compute. We compare two mathematical definitions of circuit and phenotype complexity and show how these definitions relate to other well-known properties of evolution such as redundancy, robustness, and evolvability. Using both Cartesian and Linear genetic programming implementations, we demonstrate that the logic gate circuit shares many universal properties of biologically derived G-P maps, with the exception of the relationship between one method of computing phenotypic evolvability, robustness, and complexity. Due to the inherent structure of the G-P map, including the predominance of rare phenotypes, large interconnected neutral networks, and the high mutational load of low robustness, complex phenotypes are difficult to discover using evolution. We suggest, based on this evidence, that evolving complexity is hard and we discuss computational strategies for genetic-programming-based evolution to successfully find genotypes that map to complex phenotypes in the search space.

A. H. Wright (✉) · C. L. Laue
Computer Science Department, University of Montana,Missoula, USA
e-mail: alden.wright@umontana.edu

C. L. Laue
e-mail: cheyenne.laue@mso.umt.edu

© The Author(s), under exclusive license to Springer Nature Singapore Pte Ltd. 2023
L. Trujillo et al. (eds.), *Genetic Programming Theory and Practice XIX*,
Genetic and Evolutionary Computation, https://doi.org/10.1007/978-981-19-8460-0_10

1   Introduction

One of the important questions in the study of evolution is whether there is an inherent tendency for increased Complexity in living systems. Despite the intuitive idea that organic complexity defines the world around us, relatively simple bacteria and archaea continue to be common and adaptive in many environments as well. The increase in complexity from prokaryotes to eukaryotes and from unicellular organisms to multi-cellular organisms may have been achieved through a series of major evolutionary transitions as described in [1, 2].

According to the arrow of complexity hypothesis defined by [3], evolution works to increase the complexity of already complex organisms, indicating that complexity may increase in some lineages but not in others. How then, and why, does complexity emerge as an evolutionary strategy in some cases but not others?

Further issues with understanding the evolution of complexity involves defining and measuring precisely what is meant by the term. Kolmogorov complexity is defined as the minimum length of a program that computes a specific function or string [4]. While this definition is foundational to work in a number of computational disciplines, it assesses maximum complexity to a completely random string. More recent approaches emphasize the idea that complex systems are neither completely regular nor entirely random. For example, a random string of letters, from this perspective, is no more complex than a string of letters that periodically repeats. As work by Tononi et al. [5] claims, it is functional integration in specialized systems that denotes true complexity—in the case of a string of letters, a readable piece of text is both intelligible and complex, whereas a string of completely regular or entirely random text is not.

These ideas that intelligibility, information, and environment are inherently linked in complex systems, underscores the notion that the process of evolution is fundamentally based on the transfer of information through time. In biological evolution, this information is, of course, stored in the genotypes of individuals, which map to phenotypes that in turn enable fitness-based survival and reproduction in specific environments. This correspondence, or mapping, referred to as the genotype-phenotype (G-P) map, enables phenotypes to adjust when the underlying genotypic information is altered and provides the basis for adaptive change.

While evolutionary computation models tend to simplify phenotypes to fitness functions, we claim that the field can learn from biophysical models of G-P maps. Recent work demonstrates potentially universal structural properties of biophysical and biologically-related G-P maps (see [6, 7] for a review). This paper considers the structural properties of redundancy, robustness, and evolvability, which are by definition independent of selection and fitness in our models.

Redundancy is defined in the literature as the number of genotypes that map to the same phenotype. High redundancy typically implies high robustness, which is defined as the ability to mutate a genotype without changing the phenotype that it maps to. Indeed, recent research on the structure of fitness landscapes demonstrates that those capable of realistically representing the evolution of biological processes and structures are extremely multidimensional and contain large neutral networks that connect high-fitness areas of the search space. This implies that evolution can find nearly any phenotype through neutral evolution alone, and without an associated "cost of selection" acquired through search space exploration of low-fitness areas, known as fitness valleys. Intuitively, both robustness and redundancy have significant implications for the property of evolvability, which implies the ability to evolve novel, adaptive phenotypes. The authors of [8] show that for three biologically realistic genotype-phenotype map models-RNA secondary structure, protein tertiary structure and protein complexes- even with random fitness assignment, fitness maxima can be reached from almost any other phenotype without passing through fitness valleys.

Despite the possibility of reaching any novel, adaptive form using neutral search, the reality is that complex phenotypes are both computationally and theoretically hard to evolve and a not insignificant question arises out of the research presented above—How does complexity evolve on highly redundant, neutral landscapes with corresponding high robustness, especially when highly complex phenotypes have inherently low redundancy and are both difficult to find using evolutionary search and easy to break using mutation?

In [9] and in Sect. 3.4 we define two related measures of the complexity of genotypes and phenotypes for the logic gate circuit G-P map. In comparing these definitions computationally, we found that complexity is strongly related to other structural properties of the G-P maps. Most importantly, complex phenotypes are rare, as they are represented by relatively few genotypes, and are thus hard to find relative to simpler genotypes in the search space. The phenotypic evolvability (defined in Sect. 3.4) must be approximated, which requires finding genotypes that map to the phenotype. This can be done by either random sampling or by evolution. If sampling is used, complexity is negatively related to phenotype evolvability, while if evolution is used, the relationship is positive. We show that, particularly for Cartesian Genetic Programming (CGP), evolution-based exploration of pervasive neutral networks in the genotype space is a very effective strategy for the discovery of rare/complex phenotypes.

2 The Digital Circuit G-P Map

Digital logic gate circuit G-P maps have been widely used to study the properties of evolution [10–16]. For the map used in this paper, genotypes are single-output feed-forward circuits of logic gates, such as AND and OR gates, and phenotypes are the Boolean functions computed by circuits over all possible inputs to the circuit. Phenotypes can be represented as binary strings of length 2^n where n is the number of inputs to the circuit. Thus, there are 2^{2^n} phenotypes for n-input 1-output circuits.

Our logic-gate circuits use 5 logic gates: AND, OR, NAND, NOR, and XOR. One exception is Fig. 3 which uses only the first 4 of these gates.

Table 1 shows the truth tables of these gates.

Table 1 Truth tables of logic gates

X	Y	X AND Y	X OR Y	X NAND Y	X NOR Y	X XOR Y
1	1	1	1	0	0	0
1	0	0	1	1	0	1
0	1	0	1	1	0	1
0	0	0	0	1	1	0

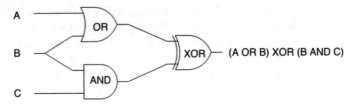

Fig. 1 Example logic-gate circuit

We compare results using two different genetic programming (GP) circuit representations: Cartesian (CGP) and Linear (LGP). The LGPvresults are new to this paper, while the CGP results are based on [9] with some results that are new to this paper.

Our CGP representation is based on [17] with one row of gates. The levels-back parameter can be chosen to be any integer from one (in which case, nodes can only connect to the previous layer) to the maximum number of nodes (in which case a node can connect to any previous node) [13]. Our LGP representation is described in [18] with the exceptions that we are using 10 instructions instead of 6, and we use the first 2 registers as computational registers and the remainder as input registers. Circuits are evaluated using bitwise instructions applied to unsigned integers.

A CGP text representation of the circuit of Fig. 1 is: `circuit((1,2,3),` `((4,OR,1,2), (5,AND,2,3), (6,XOR,4,5)))`. A gate is represented as a 4-tuple composed of the gate number, the gate function, the first input, and the second input. The last gate is the output gate. Gates are evaluated by applying the gate function to the recursively evaluated inputs to the gate. A circuit is evaluated by evaluating the output gate. Circuits process all of the possible inputs to the circuit by using the bitwise operations that are available on current computers. Thus, the state of a gate of an n-input circuit is represented as a length 2^n bit vector which is stored as an unsigned integer. Each input node of the circuit is initialized to an unsigned integer context, and these contexts are chosen so that the circuit is supplied with every possible combination of inputs. For a 2-input circuit, these are `0xc`, `0xa`, for a 3-input circuit they are `0xf0`, `0xcc`, `0xaa`, and for a 4-input circuit they are: `0xff00`, `0xf0f0`, `0xcccc`, `0xaaaa`. For an example, see Sect. 4.4.2.

An LGP representation for the same circuit is: `[(2, 1, 3, 4), (1, 2,` `4, 5), (5, 1, 1, 2)]`. A gate is represented as a 4-tuple where the elements of the tuple are the index of the gate function, the index of the output register, the index of the first input, and the index of the second input. Gates are evaluated sequentially. Computational registers are initialized to zeros and the input registers are initialized as described for the inputs of CGP circuits. Of course, the output phenotype for this example is the same as for CGP.

2.1   Neutral and Epochal Evolution

Neutral evolution is an evolutionary strategy for exploring the space of genotypes that map to a given phenotype. For each generation of the algorithm, the current genotype is randomly mutated with a point mutation which either changes the gate type or a connection between gates. If the mutated genotype maps to the same phenotype, it becomes the current genotype. Otherwise, the current genotype is unchanged. Generations continue until some is reached.

We call this strategy, which is a variant of the (1+1) evolution strategy, **neutral evolution**. This involves a random neutral walk in the network of the starting genotype and the associated phenotype. We show below that for the digital circuit G-P map that CGP neutral evolution is highly effective at exploring genotype space. This is consistent with results that show that neutral search is fundamental to the success of CGP [19].

Neutral evolution can be extended to an evolutionary strategy to find a genotype that maps to a given target phenotype. In this case neutral evolution starts from a random genotype. When the current genotype is mutated, we check to see if the mutated genotype maps to a phenotype which is Hamming distance closer to the target phenotype than the phenotype of the current genotype, and if so, we transition to neutral evolution starting with the mutated genotype. Otherwise, we continue neutral evolution based on the current genotype. The algorithm terminates when a genotype mapping to the target genotype is found or a step limit is reached. Based on [20] we call this algorithm **epochal evolution**.

These models, which do not use a population, are roughly equivalent to population-based evolution with SSWM (strong selection weak mutation) assumptions where the population is almost always isogenic. "This is due to the fact that each new mutant will either fix, replacing the genotype shared by the whole population, or become extinct before another arises. Deleterious mutations fix with such low probability that this may be assumed never to happen" [21, p. 5].

Miller and Smith [19] show that the most evolvable CGP representation is extremely large where over 95% of gates are inactive. Using CGP epochal evolution we are consistently able to evolve random 6-input 1-output phenotypes and the 7-input even parity problem by using 80 gates with 40 levels-back.

3   Genotype-Phenotype Maps

3.1   Phenotype Network and Neutral Sets

The **phenotype network** of a G-P map has phenotypes as networks and single-point mutations as edges. The **neutral set** of a phenotype is the set of genotypes that map to that phenotype. The neutral set may be path disconnected, although for the digital circuit G-P map with our parameter settings, it appears that neutral sets are either connected or have one connected component that is much larger than other components.

3.2   Redundancy

All of the G-P maps studied in the review articles Ahnert et al. [6] and Manrubia et al. [7] have the fundamental property of redundancy which is a prerequisite for the additional properties given below. **Redundancy** is the property that there are many more genotypes than phenotypes. Furthermore, the number of genotypes per phenotype varies widely from phenotype to phenotype. We show that for the CGP and LGP circuit G-P maps the frequency of genotypes per phenotypes can vary by at least 9 orders of magnitude for 3-input and 4-input circuits.

3.3   Robustness

Robustness can be defined for both genotypes and phenotypes [22]. The **robustness of a genotype** is defined as the fraction of mutations of the genotype that don't change the mapped-to phenotype. The robustness of a phenotype is the average robustness of the genotypes that map to it.

Both [6, 7] cite a linear relationship between the logarithm of the redundancy of a phenotype and its robustness as a universal property of G-P maps. We confirm this relationship for the CGP and LGP circuit G-P maps (see Fig. 4).

3.4   Evolvability

A system is evolvable if mutations can produce adaptive and heritable phenotypic variation. However, in the study of the structural properties of G-P maps, it is hard to define what is adaptive because structural properties are independent of fitness. Thus, a definition of evolvability for G-P maps should consider properties that are independent of selection. Wagner [22] defines a system as evolvable if mutations can produce heritable phenotypic variation. More specially, he defines the **evolvability of a genotype** as the number of unique phenotypes produced by mutation of the genotype. This seems to imply an antagonistic relationship between robustness and evolvability—mutations producing many unique phenotypes seems to imply a low likelihood of mutation preserving the phenotype. And in fact, this is what is almost universally observed for genotype robustness and evolvability in G-P map models.

A way around this paradox is to look at **phenotype evolvability** which Wagner defines as the number of unique phenotypes in the mutational neighborhood of the given neutral space of the phenotype. This mutational neighborhood is found by mutating all of the genotypes that map to the given phenotype. The evolvability of the phenotype is the number of unique phenotypes produced by these mutations. Both [6, 7] find that phenotype evolvability and robustness are positively related in the G-P maps that they study. However, due to the vast size of genotype spaces, the evolvability of a phenotype cannot be computed exactly except in very trivial cases. Thus, phenotype evolvability must be approximated, which necessitates finding genotypes that map to the given phenotype. Furthermore, [9] shows a negative relationship between evolvability and robustness if epochal evolution is used to invert the G-P map in the approximation of phenotype evolvability.

3.5 Universal Structural Properties

Ahnert et al. [6] reviewed structural properties for three biophysical G-P models: RNA secondary structure, the HP map of protein folding, and the polynomial model. The HP map simplifies protein folding by categorizing amino acids as either hydrophobic or hydrophilic. The polynomial model [23] is a two dimensional lattice model of self-assembly that can also be used as a G-P map to model protein quaternary structure.

After an extensive study of many biophysical G-P maps and artificial life G-P maps, Manrubia et al. [7] proposed the following: "Some of the results highlighted in the former section hint at the possibility that any sensible G-P map (and, by extension, artificial life system) is characterized by a generic set of structural properties that appear repeatedly, with small quantitative variations, regardless the specifics of each map. Extensive research performed in recent years has confirmed this possibility to an unexpected degree.

"Some of the commonalities documented are navigability, as reflected in the ubiquitous existence of large neutral networks for common phenotypes that span the whole space of genotypes, a negative correlation between genotypic evolvability and genotypic robustness, a positive correlation between phenotypic evolvability and phenotypic robustness, a linear growth of phenotypic robustness with the logarithm of the NSS [neutral set size], or a near lognormal distribution of the latter" [7].

We investigate these properties in the context of our CGP and LGP circuit models. We confirm these properties with one significant and surprising exception in regards to the relationship between robustness and evolvability which we explain below.

4 Structural and Complexity Properties for the Circuit G-P Map

4.1 Redundancy and Bias

In this section we show the extreme variation in the frequency of digital circuit phenotypes. Figure 2 shows the frequencies of all $2^{2^4} = 65536$ phenotypes ordered from most frequent to least frequent. The CGP frequencies are based on a sample of 10^{10} genotypes and the LGP frequencies are based on a sample of 2×10^{11} genotypes. Despite the much larger sample, rare phenotypes are much better represented using CGP then using LGP. There are 531 unrepresented phenotypes for LGP and none for CGP.

The choice of gate functions contributes to the number of phenotypes discovered by sampling. While all of the other plots in this paper are generated using the 5 gate functions AND, OR, NAND, NOR, and XOR, Fig. 3 uses only the first 4 of these gates. In this case only about 1/3 of phenotypes for LGP and 2/3 for CGP despite the 20 times larger sample for LGP.

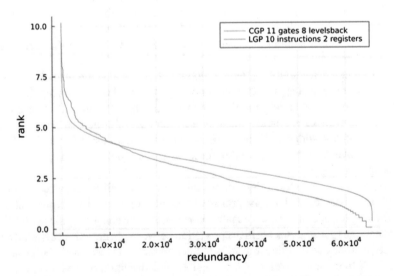

Fig. 2 Log redundancy versus rank by sampling with XOR gate. CGP: 10^{10} samples, LGP: 2×10^{11} samples

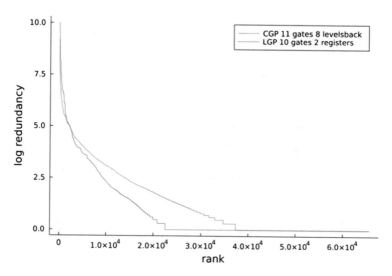

Fig. 3 Log redundancy versus rank sampling without XOR gate. CGP: 10^{10} samples, LGP: 2×10^{11} samples

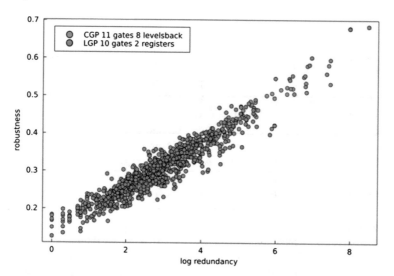

Fig. 4 Robustness versus log redundancy CGP 11 gates 8 lb LGP 10 instructions

4.2 Robustness

Figure 4 confirms the strong linear relationship between robustness and log redundancy which is universal in the G-P maps reviewed by [6].

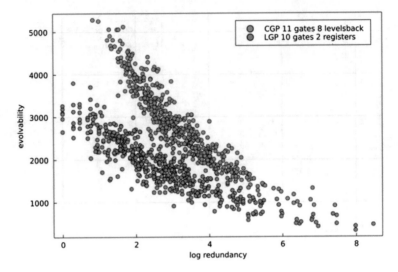

Fig. 5 Evolution evolvability versus Log redundancy CGP 11 gates 8 lb LGP 10 instructions

4.3   Evolvability

We approximated the evolvability of a phenotype by counting the total number of unique phenotypes in the mutational neighborhoods of 600 genotypes. These geno-types were found by using epochal evolution with the phenotype as the target [9] refers to this method as **evolution evolvability**. An alternative method, sampling evolvability, is based on using a sampling methodology to find genotypes that map to the given phenotype is described in [9]. Reference [18] uses random walks to implement a sampling methodology. As shown in Fig. 7 of [9], evolution evolv-ability is strongly negatively related to log robustness while sampling evolvability is strongly positively related to log robustness. The negative relationship between log redundancy and evolvability is different from the biologically related G-P maps described in [6, 7].

Figure 5 shows a strong negative relationship between log redundancy and pheno-type evolvability for 500 random 4-input phenotypes. The same random phenotypes were used for CGP and LGP. These results show that evolution evolvability is much less LGP than for CGP.

Figure 6 shows a strong negative relationship between evolution phenotype evolv-ability and robustness. Again, LGP evolvability is much less than CGP evolvability.

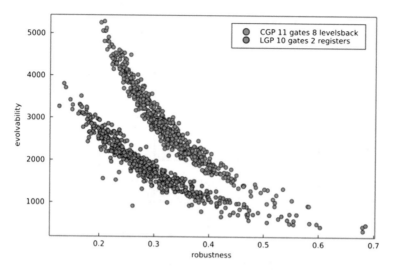

Fig. 6 Evolution evolvability versus robustness CGP 11 gates 8 lb LGP 10 instructions

4.4 Tononi and Kolmogorov Complexity

We use two methods to compute the complexity of the digital circuits representing phenotypes in our model. First, we apply the Tononi measure of complexity defined for neural circuits in [5].

4.4.1 Entropy and Mutual Information

Based on work by [5, 24] the entropy $H(X)$ of a discrete random variable X with N states is given by

$$H(X) = -\sum_{i=1}^{N} p_i \log(p_i)$$

where p_i is the probability of state i. The entropy of X is highest when all states of X are distinct and have equal probability. It is lowest when X has only one state.

The mutual information of two subsets A and B of a system X is given by

$$MI(A; B) = H(A) + H(B) - H(X)$$

If A and B are independent, then $MI(A; B) = 0$, while if A can be completely predicted knowing B, then $MI(A; B) = H(A) = H(B) = H(X)$. Mutual information has the advantages that it is multivariate and captures both linear and non-linear dependencies.

The uncertainty about the state of a subset A of the system X which is accounted for by the state of the rest of the system $X - A$ is given by

$$MI(A; X - A) = H(A) + H(X - A) - H(X)$$

[5, Equation 2].

4.4.2   Tononi Complexity

Given a circuit, let X be the binary matrix with rows corresponding to gates and columns corresponding to the 2^n possible inputs to the circuit. As described in Sect. 2 the inputs to an n-input circuit are specified by the n contexts which are the same for any n-input circuit. Each row of X is the state of the corresponding gate when the circuit is executed.

Following is the complete state of the circuit of Fig. 1 The first 3 rows correspond to the 3 standard contexts for a 3-input circuit, and the matrix X is the last 3 rows. The hexadecimal representation of each row is given as the last column. The last row represents the output phenotype of the circuit.

Input 1	1	1	1	1	0	0	0	0	0xf0
Input 2	1	1	0	0	1	1	0	0	0xcc
Input 3	1	0	1	0	1	0	1	0	0xaa
Gate 4 OR	1	1	1	1	1	1	0	0	0xfc
Gate 5 AND	1	0	0	0	1	0	0	0	0x88
Gate 6 XOR	0	1	1	1	0	1	0	0	0x74

We compute the entropy of X by interpreting X as a probability distribution over the columns of X. Each column is a binary vector which can be interpreted as a bit string. For the above example, there are two occurrences of 110, four occurrences of 101, and two occurrences of 000. Thus, the probability of 110 is 1/4, the probability of 101 is 1/2, and the probability of 000 is 1/4. We find that the entropy $H(X)$ is given by $H(X) = -1/4 \log(1/4) - 1/2 \log(1/2) - 1/4 \log(1/4) = 3/2$. (We use base 2 logarithms.)

Following [5, 24], we will use X_j^k to denote a sub-matrix of X of cardinality k where j is an index over different subsets of cardinality k. Continuing our example, let $X\{1, 3\}$ denote the sub-matrix of X corresponding to rows 1 and 3 of X. Thus, $X\{1, 3\}$ would be X_j^2 for some j. We denote the entropy of $X\{1, 3\}$ by $H\{1, 3\}$

$$X\{1, 3\} = \begin{vmatrix} 1\ 1\ 1\ 1\ 1\ 1\ 0\ 0 \\ 0\ 1\ 1\ 1\ 0\ 1\ 0\ 0 \end{vmatrix}$$

This matrix has columns 10, 11, and 00 with probabilities 1/4, 1/2, and 1/4 respectively. Thus, $H(A) = H\{1, 3\} = 3/2$.

If M is the number of gates, there are $\binom{M}{k} = \frac{M!}{k!(M-k)!}$ ways of choosing X_j^k.

Tononi's definition [5, 24] of the complexity $C(X)$ of a circuit whose matrix is X is given by

$$C(X) = \sum_{k=1}^{M/2} < MI(X_j^k; X - X_j^k) > = \frac{1}{2} \sum_{k=1}^{M} \frac{1}{\binom{M}{k}} \sum_{j=1}^{\binom{M}{k}} MI(X_j^k; X - X_j^k) \quad (1)$$

where <> denotes ensemble average (the average over j). The papers [5, 24] give alternative equivalent information theoretic formulas for complexity.

To continue the example, the complexity of the circuit of Fig. 1 is computed as:

$$
\begin{aligned}
C(X) &= MI(X\{1\}; X\{2, 3\}) + MI(X\{2\}; X\{1, 3\}) + MI(X\{3\}; X\{1, 2\}) \\
&= 1/3((H\{1\} + H\{2, 3\} - H\{1, 2, 3\}) + (H\{2\} + H\{1, 3\} - H\{1, 2, 3\}) \\
&+ (H\{3\} + H\{2, 3\} - H\{1, 2, 3\}) \\
&= 1/3(0.8113 + 0.8113 + 1.0) \\
&= 0.8742
\end{aligned}
$$

The Tononi complexity of a digital circuit phenotype is the average of the Tononi complexity of a sample of circuits that map to the phenotype.

4.4.3 Kolmogorov Complexity

The second method we use to compute complexity is a version of Kolmogorov complexity designed for digital circuits. The Kolmogorov complexity of a string is defined as the minimum length of a computer program that generates the string [25].

Digital circuit phenotypes can be expressed as bit strings, and computer circuits are a language for generating these strings. Based on this logic, we define the Kolmogorov complexity of a digital circuit phenotype as the minimum number of logic gates needed in a circuit to map to a specified phenotype. This assumes a specific representation (CGP or LGP) with a fixed levels-back or number of registers. As a simple example, there is no single gate CGP circuit that computes the function for the phenotype A EQV B (bit string 1001 = hex 0x9). However, since the CGP circuit ((1,2), ((3, XOR, 2,1), (4,NOR,3,3))) computes this phenotype, there is a 2-gate circuit that also computes the function. Here we see that the Kolmogorov complexity of this phenotype is 2.

Figure 7 shows that Tononi and Kolmogorov complexity are empirically consistent for one setting of CGP parameters. This plot is taken from Fig. 3 of [9].

Ahnert et al. [26] define a complexity for the polynomial G-P map which is based on Kolmogorov complexity.

Dingle et al. [27] show that evolving genotypes to phenotypes of high Kolmogorov complexity is computationally hard. The authors consider computable maps of the form $f : I \rightarrow O$ where I is a collection of input sequences and O is the corre-

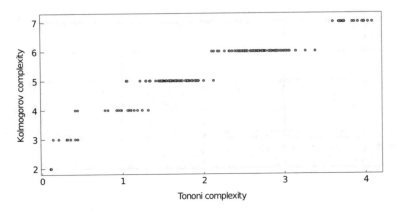

Fig. 7 The relationship between Tononi and Kolmogorov complexity CGP 4 inputs 11 gates 8 levels-back

sponding collection of outputs which can be described as discrete sequences. When applied to our definition of Kolmogorov complexity for digital circuits, I is the standard context for the number of inputs, and f is the program that evaluates the circuit as a function of the context. This program is part of our simulation code and does not change as the size of the input sequences increases.

If $x \in O$, Dingle et all defined $P(x)$ to be the probability that a random input from I will give will map to x. Their main result is their Eq. 3:

$$P(x) \leq 2^{-a\tilde{K}(x)-b}$$

where $\tilde{K}$ is an approximate Kolmogorov complexity and the constants $a \geq 0$ and b depend on the mapping f and not on x. The result is based on the assumption that f is of limited complexity, which is demonstrated for digital circuits as the code for f does not change as the input size increases.

It appears that the Dingle et al. result applies to digital circuit G-P maps in which case evolving complexity is hard from a computational complexity point of view in a way that scales to all circuit input sizes. Our simulation results confirm this theoretical prediction. An application of the Dingle et al. [27] theoretical result to G-P maps is [28].

4.5 Complexity Density

Figures 8 and 9 show that the Tononi and Kolmogorov complexity of random CGP genotypes is much less than the complexity of random phenotypes. The Tononi complexity of phenotypes is computed by evolving genotypes that map to a phenotype,

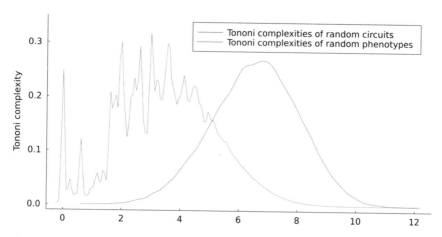

Fig. 8 Complexity densities of genotypes and phenotypes CGP 4 inputs 11 gates 8 levels-back

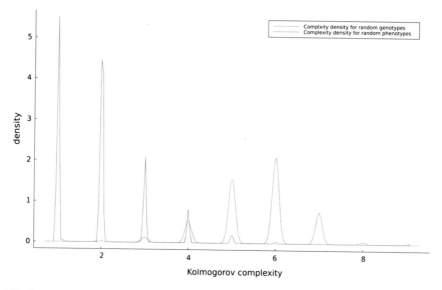

Fig. 9 Kolmogorov complexity densities of genotypes and phenotypes CGP 4 inputs 11 gates 8 levels-back

which may give an increased estimate of complexity in comparison to sampling. Figure 8 is taken from Fig. 5 of [9].

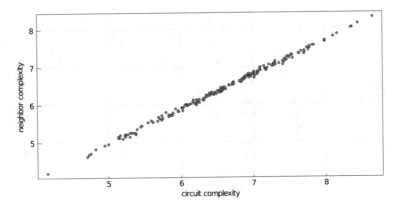

Fig. 10 Average complexity of neighboring genotypes CGP 4 inputs 11 gates 8 levels-back

4.6 Complexity Correlations

Figure 10 shows that the average Tononi complexity of 1-mutant neighbors of a genotype is close to the Tononi complexity of the genotype. This figure is taken from Fig. 4 of [9]

4.7 Complexity and Redundancy

4.7.1 Complex Phenotypes Are Rare

Figure 11 shows a strong negative relationship between Tononi complexity and log redundancy. Note that the computation of Tononi complexity for CGP differs from that of LGP which is an explanation for the larger values of Tononi complexity for LGP. The same 500 random phenotypes were used for these plots as for the evolvability plots.

For both representations, complex phenotypes are rare, i. e., they are represented by a small number of genotypes, indicating that they are hard to evolve.

4.8 Complexity and Robustness

4.8.1 Complex Phenotypes Have Low Robustness

Figure 12 shows the strong negative relationship between robustness and Tononi complexity. The plot is very similar to Fig. 11 which is not surprising given the strong positive relationship between log redundancy and robustness shown in Fig. 4.

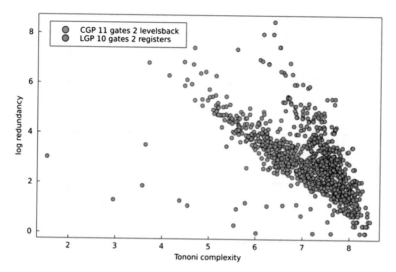

Fig. 11 Log redundancy versus complexity CGP 11 gates 8 lb LGP 10 instructions

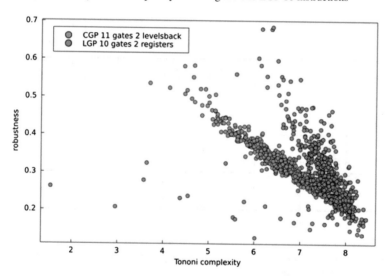

Fig. 12 Robustness versus complexity CGP 11 gates 8 lb LGP 10 instructions

4.9 Complexity and Evolvability

Figure 13 shows that evolution evolvability increases with Tononi complexity for both CGP and LGP. This confirms the result of Fig. 8a of [9]. Note that sampling evolvability has a negative relationship with complexity as shown in Fig. 8b of [9].

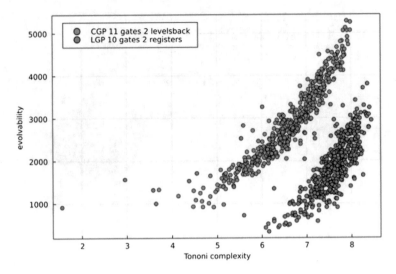

Fig. 13 Evolution evolvability versus complexity CGP 11 gates 8 lb LGP 10 instructions

The positive relationship between evolution evolvability and complexity suggests that as complexity evolves, a greater number of new phenotypes are nearby, and thus there is a positive feedback enabling the discovery of further complexity.

5   Conclusion

The review article [7] hypothesizes that the evolutionary process is often more determined by the structure of the relevant G-P map than by selection. Two examples are the "survival of the flattest" phenomenon where genotypes with high robustness (i. e., low mutational load) will out-compete genotypes with higher fitness but lower robustness [29], and the "arrival of the frequent" phenomenon where infrequent (rare) phenotypes are not discovered in time to compete with frequent (common) phenotypes [30]. Thus, understanding the structural properties of G-P maps is important for understanding evolution.

We have compared CGP and LGP representations of digital logic gate circuit G-P maps, where genotypes are circuits and phenotypes are the functions computed by circuits. We have shown that both representations have nearly all of the structural properties proposed as universal for biologically related and inspired G-P maps. Our models show that redundancy varies by many orders of magnitude between phenotypes and in CGP models, even those containing rare (low redundancy) phenotypes, neutral sets tend to percolate through genotype space, and this is true although to a lesser extent) for LGP. Thus, exploring neutral sets is an effective strategy for evolving genotypes (circuits) that map to a given phenotype.

Neutral evolution is enabled by robustness which is the fraction of mutations of a genotype which do not change the mapped-to phenotype. The evolvability of a phenotype is the number of unique phenotypes which are one mutation away from a genotype in the neutral set of the phenotype. We found a negative relationship between evolvability approximated using epochal evolution and robustness, which differs from results found in biologically-related G-P maps. On the other hand, if evolvability is approximated by using a large random sample of genotypes we show this relationship changes from negative to positive (which is the "universal" result for biologically related G-P maps). This was shown for CGP in our previous paper [9].

We defined an information-theoretic measure of circuit (genotype) complexity based on [5] that we call Tononi complexity, and we defined Kolmogorov complexity for phenotypes based on the minimum number of gates needed to compute a phenotype. Furthermore, we empirically demonstrated that these two measures of complexity are related.

Why then is evolving genotypes that map to complex phenotypes so difficult? First, random genotypes have low complexity as shown in Figs. 8 and 9. Second, complex phenotypes tend to have both low redundancy and low robustness as shown in Figs. 11 and 12. Third, the computational complexity results of [27] show that the probability of sampling genotypes that map to a phenotype decreases exponentially with the Kolomogorov complexity of the phenotype.

How then does complexity evolve at all? Our results indicate some strategies that facilitate the evolution of complex phenotypes.

To begin, the average complexity of the genotypes near to a given phenotype tends to be close to the complexity of a given genotype as shown in Fig. 10 Based on this, conducting evolutionary searches near complex phenotypes should lead to genotypes of high complexity as well. Second, our results demonstrating a positive relationship between complexity and evolution evolvability: see Fig. 13. suggest that as genotypes evolve those that map to neutral sets will be mutationally accessible from the genotypes of many different phenotypes.

5.1 Further Work

1. Determine whether the circuit G-P maps have the shape-space covering property. A G-P map has the shape space covering property that, if given a phenotype, only a small radius around a genotype encoding that phenotype needs to be explored in order to find the most common phenotypes.
2. Find additional computational problems whose G-P map has the universal structural properties described above.
3. Investigate the reason that genotypes evolved to map to a target phenotype have higher Tononi complexity than genotypes sampled to map to the target phenotype.

Acknowledgements The input of GPTP reviewers Stu Card and Ting Hu was very helpful. Cooper Craig produced some of the plots. And we thank Jesse Johnson for the use of computational resources.

References

1. Szathmáry, E., Smith, J.M.: The major evolutionary transitions. Nature **374**(6519), 227–232 (1995)
2. Szathmáry, E., Smith, J.M.: The Major Transitions in Evolution. WH Freeman Spektrum Oxford, UK (1995)
3. Bedau, M.A.: The evolution of complexity. In: Mapping the Future of Biology, pp. 111–130. Springer (2009)
4. Li, M., Vitányi, P., et al.: An Introduction to Kolmogorov Complexity and Its Applications, vol. 3. Springer (2008)
5. Tononi, G., Edelman, G.M., Sporns, O.: Complexity and coherency: integrating information in the brain. Trends Cognit. Sci. **2**(12), 474–484 (1998)
6. Ahnert, S.E.: Structural properties of genotype-phenotype maps. J. R. Soc. Interface **14**(132), 20170275 (2017)
7. Manrubia, S., Cuesta, J.A., Aguirre, J., Ahnert, S.E., Altenberg, L., Cano, A.V., Catalán, P., Diaz-Uriarte, R., Elena, S.F., García-Martín, J.A., et al.: From genotypes to organisms: state-of-the-art and perspectives of a cornerstone in evolutionary dynamics. Phys Life Rev **38**, 55–106 (2021)
8. Greenbury, S.F., Louis, A.A., Ahnert, S.E.: The structure of genotype-phenotype maps makes fitness landscapes navigable. bioRxiv (2021)
9. Wright, A.H., Laue, C.L.: Evolvability and complexity properties of the digital circuit genotype-phenotype map. In: Proceedings of the Genetic and Evolutionary Computation Conference, pp. 840–848 (2021)
10. Ofria, C., Wilke, C.O.: Avida: evolution experiments with. In: Artificial Life Models in Software, p. 1 (2005)
11. Arthur, W.B., Polak, W.: The evolution of technology within a simple computer model. Complexity **11**(5), 23–31 (2006)
12. Macia, J., Solé, R.V.: Distributed robustness in cellular networks: insights from synthetic evolved circuits. J. R. Soc. Interface **6**(33), 393–400 (2009)
13. Miller, J.F., Harding, S.L.: Cartesian genetic programming. In: Proceedings of the 11th Annual Conference Companion on Genetic and Evolutionary Computation Conference: Late Breaking Papers, pp. 3489–3512 (2009)
14. Raman, K., Wagner, A.: The evolvability of programmable hardware. J. R. Soc. Interface **8**(55), 269–281 (2011)
15. Hu, T., Payne, J.L., Banzhaf, W., Moore, J.H.: Evolutionary dynamics on multiple scales: a quantitative analysis of the interplay between genotype, phenotype, and fitness in linear genetic programming. Genet. Program. Evol. Mach. **13**(3), 305–337 (2012)
16. Hu, T., Banzhaf, W.: Neutrality, robustness, and evolvability in genetic programming. In: Genetic Programming Theory and Practice XIV, pp. 101–117. Springer (2018)
17. Miller, J.F., Job, D., Vassilev, V.K.: Principles in the evolutionary design of digital circuits-Part I. Genet. Program. Evol. Mach. **1**(1–2), 7–35 (2000)
18. Hu, T., Tomassini, M., Banzhaf, W.: A network perspective on genotype-phenotype mapping in genetic programming. Genet. Program. Evol. Mach. 1–23 (2020)
19. Miller, J.F., Smith, S.L.: Redundancy and computational efficiency in cartesian genetic programming. IEEE Trans. Evolut. Comput. **10**(2), 167–174 (2006)
20. Crutchfield, J.P., van Nimwegen, E.: The evolutionary unfolding of complexity. In: Landweber, L.F., Winfree, E. (eds.) Evolution as Computation. Natural Computing Series, pp. 67–94. Springer Berlin (2002)
21. Nichol, D., Robertson-Tessi, M., Anderson, A.R.A., Jeavons, P.: Model genotype-phenotype mappings and the algorithmic structure of evolution. J. R. Soc. Interface **16**(160), 20190332 (2019)
22. Wagner, A.: Robustness and evolvability: a paradox resolved. Proc. R. Soc. B: Biol. Sci. **275**(1630), 91–100 (2008)

23. Greenbury, S.F., Johnston, I.G., Louis, A.A., Ahnert, S.E.: A tractable genotype-phenotype map modelling the self-assembly of protein quaternary structure. J. R. Soc. Interface **11**(95), 20140249 (2014)
24. Tononi, G., Sporns, O., Edelman, G.M.: A measure for brain complexity: relating functional segregation and integration in the nervous system. Proc. Natl. Acad. Sci. **91**(11), 5033–5037 (1994)
25. Li, M., Vitanyi, P.M.B.: Kolmogorov Complexity and Its Applications. Centre for Mathematics and Computer Science (1989)
26. Ahnert, S.E., Johnston, I.G., Fink, T.M.A., Doye, J.P.K., Louis, A.A.: Self-assembly, modularity, and physical complexity. Phys. Rev. E **82**(2), 026117 (2010)
27. Dingle, K., Camargo, C.Q., Louis, A.A.: Input-output maps are strongly biased towards simple outputs. Nat. Commun. **9**(1), 1–7 (2018)
28. Johnston, I.G., Dingle, K., Greenbury, S.F., Camargo, C.Q., Doye, J.P.K., Ahnert, S.E., Louis, A.A.: Symmetry and simplicity spontaneously emerge from the algorithmic nature of evolution. Proc. Natl. Acad. Sci. **119**(11), e2113883119 (2022)
29. Wilke, C.Q., Wang, J.L., Ofria, C., Lenski, R.E., Adami, C.: Evolution of digital organisms at high mutation rates leads to survival of the flattest. Nature **412**(6844), 331–333 (2001)
30. Schaper, S., Louis, A.A.: The arrival of the frequent: how bias in genotype-phenotype maps can steer populations to local optima. PloS one **9**(2), e86635 (2014)

ESSAY: Computers Are Useless ... They Only Give Us Answers

Bill Worzel

1 Introduction

While the title of this essay, taken from a quotation from Picasso [1], is somewhat facetious, it contains more than a grain of truth. We have all experienced results that left us scratching our heads with Evolutionary Algorithms, but often these algorithms lead us to the most fruitful results. It can be hard to create the right conditions to find good answers. To quote Shakespeare: "I can call the spirits from the vasty deep!" to which the response in the text is "Why so can I! And so can any [person], but will they come when you doth call?"

In the past, I have usually spent most of my time using GP to solve a particular problem. In this essay, however, I want to venture into the murky depths of theory, or at least the larger world of what we could learn from the *computing singularity* (not to be confused with astronomical singularities). For those who are not familiar with the concept of the *computing singularity*, it is the idea that a time will come when a computer will be truly smarter than humans. But where do we start?

At the first workshop on Genetic Programming—Theory and Practice (GPTP I) [2], John Holland in his keynote speech pointed out that "Theory tells us where to look!" but he followed this up by adding that "Practice tells us when we are looking in the wrong place." However, that doesn't say where the right place is! It would be easy to say that the right place is everywhere that isn't where we are looking—but that can be a mighty big place.

B. Worzel (✉)
Evolution Enterprise, Ann Arbor, MI, USA
e-mail: billwzel@gmail.com

© The Author(s), under exclusive license to Springer Nature Singapore Pte Ltd. 2023
L. Trujillo et al. (eds.), *Genetic Programming Theory and Practice XIX*,
Genetic and Evolutionary Computation, https://doi.org/10.1007/978-981-19-8460-0_11

2 Recommended Sources

Here, I hope to at least start to find my path to the Singularity. I shall mention a few important books that have influenced my inquiry and that I highly recommend to others.

2.1 Hidden Order

To start, I have used as my texts the John Holland book "Hidden Order" [4], which helps to define complexity, and his book "Emergence—From Chaos to Order" [5]. It was John Holland's work [3] that led John Koza to find the building blocks for genetic programming. I believe that the GP community is more than aware of these books [3–5], so I will not go into a discussion of details, but only touch them briefly.

In John Holland's book "Hidden Order" [4], he laid out the importance of finding the building blocks for adaptive systems in pursuit of a general system in a specific realm. No matter what realm you are talking about, you need to know what these blocks are, and why these are the ones you are looking for, otherwise you won't recognize the result when you get it!

We can use computers to simplify things that we know how to do, but it is much harder to make new discoveries. In my nearly 20 years of using genetic programming, I personally have only once or twice felt that I had truly discovered something that was novel and surprising. The rest of the time, the only surprise was if a numerical value or result was outside of a range of solutions. Pablo Picasso divined that finding the question and, even better, the unexpected is what we really should hope for.

2.2 Gödel, Escher, Bach

Douglas Hofstadter's "Gödel et al." [6] and its complementary book "Metamagical Themas" [7] were what ultimately inspired Melanie Mitchell's new book "Artificial Intelligence, A Guide for Thinking Humans" [8] which goes on the list as it is, perhaps, the most up-to-date book on AI of those that I have recommended here.

2.3 The Cambridge Quintet

I would also want to add "The Cambridge Quintet," written by Casti [9]. He calls it "a work of scientific speculation" and a non-technical novel titled "Machinehood" which was written by Divya [10]. Though "Machinehood" is a work of fiction, such books often suggest new paths of thought. It supposes that a machine becomes fully sentient and depicts commonplace interactions between the machine and robots. Finally, I want to add the book "The Mangle of Practice" by Pickering [11] which, while not entirely focused on computers per se, focuses on how people interact with machines in general, and robotic machines specifically.

I will take these one-by-one and describe why I think they have something to say about the Singularity. First though: How does one define the Singularity? In Melanie Mitchell's book, she cites Ray Kurzweil as defining the AI Singularity as "... a time in the near future when computers will become smarter than humans." I will take this as an operational definition for the purposes of this essay.

The Cambridge Quintet takes the concept of a number of historical luminaries debating whether a machine could become sentient. In this guise, C.P. Snow has been asked to report to the English government whether "a computing machine could be as good as a human in its cognitive capacity?" and "Even more generally: Could a machine ever be developed to the extent that we would accord it full human rights?" The other "participants" in this debate are Erwin Schrödinger of quantum physics fame, Alan Turing of the Turing Test, JBS Haldane who was one of the early scientists to combine genetics and evolutionary theory, and the philosopher Ludwig Wittgenstein. Casti hypothetically brings them all to Cambridge to debate whether a computer could become sentient.

While The Cambridge Quintet is not definitively on a singularity, in their hypothetical debate, it raises a number of ideas that are quite relevant to our hunt for the Singularity. Moreover, since I intend to compare books that touch on the Singularity, I shall put this into the mix. Some of the ideas debated in this book are as follows:

- Can computers understand human language?
- Can computers replicate themselves? And if so, can the replication create an "offspring" computer that can differ from its parent(s)?
- Human thinking requires complicated mental states and the manipulation of symbols. Can this ability be matched by computers and if so, would it mirror human thinking? Is it possible that computers could truly think differently than people?
- Could a computer understand fiction and know that it is fiction?
- Could a computer be able to describe something it does not know? (e.g., "It looked like a ball but it floated into the sky and had a string.") Could the computer recognize it as a child's balloon?
- Could a computer understand the context; eg, could it remember seeing a picture of a balloon and realize that it is a balloon?
- Could a computer understand Art? Could it understand Abstract Art?

There are certainly many more tests one could propose.

3 Artificial Intelligence

The next book considered is Melanie Mitchell's book "Artificial Intelligence" [8]. Melanie Mitchell's subtitle is "A Guide for Thinking Humans," which suggests that we aren't in Kansas anymore :)! It is a relatively new book and Mitchell is a top writer and explainer of Artificial Intelligence, though in this case, AI means deep learning. Her prologue is titled "Terrified." But it is not Melanie Mitchell who is terrified, but Douglas Hofstadter who is terrified when Google invites them to give them their view of the coming age of AI. Hofstadter has growing doubts about the possibility of AI replacing humans. Not only is he worried about replacing people but he worries that AI would become too smart and too creative to the point that we humans would become superfluous.

From there, Mitchell gives a good but relatively short chapter on the birth of the AI Perceptrons that lead to the formalism of neural networks. She touches on the early developers and the development of error backpropagation for neural nets. She also touches on how many new uses of neural nets are so commonplace that most people do not know that they are using AI.

Then the book digs into how deep learning is used for recognizing images using trained neural nets, in games, including—but not limited to—chess, and in general classification tasks. It proceeds to robotics and perhaps shows the limits of deep learning for images, machine translation, game playing computers (again), and so-called "Beneficial AI" such as healthcare for diagnostics, prognostics, and other uses. So what does all this have to do with GP? GP has been used in many of the same applications that Deep Learning has been used. For some purposes, GP is better, and for others, deep learning is better. For example, for diagnostics and prognostics, GP is better because its results and how they were derived at are more transparent than those of neural nets. All the while deep learning is better suited for visual uses and which is better than the other in which situation is left to the reader.[1]

4   The Mangle of Practice, Times, Agency, and Science

This book by Pickering [11] is something that does not focus only on machines but the practice of tools, where tools would include concepts on tool use. This notion is not at all foreign for computer programmers as we use code to build new tools and there are few programmers that don't use libraries or at least their own custom tools. Even when we have custom tools, the act of programming is creating a new tool.

This book puts the idea of tool making into the practice of science and its methods. In modern parlance, it is a "meta-book." It does this by deconstructing the tools (where "tools" include the practice of the use of tools) of science and examines their uses, including novel uses of well-known tools such as scanning tunneling microscopes.

---

[1] A deep learning bibliography can be found here: https://scholar.google.com/scholar?q=deep+learning+bibliography&hl=en&as_sdt=0&as_vis=1&oi=scholart.

5 Combinators

Finally, Stephen Wolfram's book "Combinators, A Centennial View" shows another side of computing [13]. In his preface, he points out that combinators predate what we consider "normal computing." Wolfram notes that "Combinators were used before Turing machines, before Lambda calculus—even before Gödel's theorem there were combinators. They were the very first abstract examples ever to be constructed of what we now know as universal computation—and they were first presented on December 7, 1920." Why would we want to involve combinators in GP? After all, we have a perfectly good way of developing results using GP, so why bring in something that is different from the usual procedural computing that we use every day? As I said at the workshop this year, combinators have some unique characteristics. In particular, they are agglutinative. It is not unusual to see things like s[s[k[s]][s[k[s[k[f]]]][g]]][k[s[k][k]][x][y]—which reduces to f[g[x][y]][y]].

Because the density of the code is not easy to write (though people do use functional programming for computing). I believe that it is better suited as a good substrate for AI because functional code can be difficult if you are not fluent in its use. For these reasons, I intend to explore it as a machine learning language. Previously Duncan MacLean and I have included functional programming in combination with "normal" programming in C and C++ [12] and I plan to use pure functional programming in GP. I believe that this is something worth exploring and it may be a powerful way to jump-start AI.

6 Where Next?

As we proceed toward the computing Singularity, I believe that functional languages will have an edge in building machine learning for the reasons mentioned above: the agglutinative capability and the simplicity of computer hardware built on combinators. Indeed, Stephen Wolfram calls combinators the "Ultimate Symbolic Abstraction" and notes that there have been hardware systems built that are specifically optimized for combinators which are actually much simpler than "normal" computers and have natural properties that make parallel processing easier. Based on work done previously in Cambridge, UK, a computer was built in the 1970s with a handful of TTL (transistor-to-transistor logic). This became the first combinator-based computer in the modern era.

Now that we have powerful custom-made chips, it is easy to make chips for combinators. With this ability, we can build computers using combinators and other functional computers. Using GP with pure combinator computing, we can use Wolfram's book as an instruction manual for a new way of developing code, including in GP.

7   Summary

We are at the cusp of a new revolution in computing. We already are seeing a new paradigm in computing with AI, but this is the tip of the iceberg. The combination of GP and deep learning will be leading the charge. While deep learning has already gone to the forefront of this sea change, GP has yet to be recognized for its potential role in this development. The combination of these tools and tools that are yet to be developed will give us this profound change and perhaps even the Singularity.

References

1. Picasso, P.: https://www.artlyst.com/news/picasso-denounced-computers-in (1968)
2. Riolo, R., Worzel, B. (eds.): Genetic Programming Theory and Practice. Springer, US (2003)
3. Holland, J.H.: Adaptation in Natural and Artificial Systems. The MIT Press (1992)
4. Holland, J.H.: Hidden Order, p. 208. Basic Books (1996)
5. Holland, J.H.: Emergence: From Chaos to Order, p. 777. Basic Books (1979)
6. Hofstadter, D.R., Gödel, E.: Bach: an Eternal Golden Braid, 272. Basic Books (1999)
7. Hofstadter, D.R.: Metamagical Themas: Questing for the Essence of Mind and Pattern, p. 852. Basic Books (1985)
8. Mitchell, M.: Artificial Intelligence: A Guide for Thinking Humans, p. 313. Farrar Straus and Guroux, New York (2019)
9. Casti, J.L., The Cambridge Quintet: A Work Of Scientific Speculation, p. 208. Basic Books (1999)
10. Divya, S.B.: Machinehood, p. 416. Gallery/Saga Press (2021)
11. Pickering, A.: The Mangle of Practice: Time, Agency, and Science, p. 296. University of Chicago Press (1995)
12. Worzel, W.P., MacLean, D.: In: Riolo, R., Worzel, W., Kotanchek, M. (eds.) Genetic Programming Theory and Practice XII, pp. 53–72. Springer, Cham (2015)
13. Wolfram, S.: Combinators: A Centennial View, p. 852. Wolfram Media (2021)

Index

A

Abductive reasoning, 142, 151
Adversarial model, 118, 119, 121, 137
AI Planning, 58
Artificial dorsal stream, 154
Artificial intelligence, 102, 257
Artificial neural networks, 6, 120, 172, 258
Artificial ventral stream, 153
Atomic potentials, 3
Automated machine learning, 202
Automatic programming, 32, 143
Autopoiesis, 182

B

Benchmark, 35, 65, 145
Bias, 147, 203–205
Brain programming, 153

C

Cartesian genetic programming, 235
Catastrophic forgetting, 147, 148
Circuit, 172, 235
Classification, 32, 85, 147, 175, 258
Code re-use, 145
Co-evolution, 78, 118, 121
Complexity, 2, 84, 93, 145, 168, 203, 234, 256
Contingent, 143, 169
Correlation, 12, 33, 34, 84, 96, 162, 206, 239

D

Data-driven technology, 148
Data science, 202

Deductive reasoning, 146
Development, 58, 156, 169
Domain knowledge, 89, 115, 144

E

Embodiment, 168
Evolutionary art, 117
Evolutionary generative adversarial model, 118, 119, 121, 128, 130, 133
Evolvability, 119, 124, 126, 128, 132, 133, 234
Explainability, 82, 148

F

Feature importance, 82
Feature selection, 84, 202
Fibonacci sequence, 162, 164
Fitness function, 6, 33, 34, 59, 120, 123–125, 128–130, 145, 234

G

Generalization, 12, 95, 145, 213
Generative adversarial network, 121, 122
Genetic programming, 58, 84, 92, 118, 142, 202, 255
Genomics, 92
Genotype, 234
Genotype-phenotype mapping, 120, 122
GPU, 118, 122, 124, 147

H

Hierarchical architectures, 148

© The Editor(s) (if applicable) and The Author(s), under exclusive license to Springer
Nature Singapore Pte Ltd. 2023
L. Trujillo et al. (eds.), *Genetic Programming Theory and Practice XIX*,
Genetic and Evolutionary Computation, https://doi.org/10.1007/978-981-19-8460-0

Hierarchical task network, 63
Human intelligence, 157

I
Image classification, 142
Image generation, 117–121, 132, 137
Inductive reasoning, 150
Inferential knowledge, 142
Interpretability, 6, 82, 203, 204

K
Kolmogorov complexity, 234

L
Linear genetic programming, 86
Logic gate, 173, 235
Loss function, 33

M
Machine learning, 2, 32, 81, 94, 142, 202,
       259
Materials science, 2
Metabolomics, 82, 92
Meta-dynamics, 190
Metamaterials, 189
Missing data, 94
Modeling, 2, 94, 143, 202
Modularity, 145
Mutual information, 115, 173, 243

O
Object detection, 142

P
Pareto front, 84, 96
Pathfinding, 60
Performance metrics, 213
Phenotype, 86, 234
Potential energy surfaces, 2

Prediction, 4, 13, 81, 84, 93, 175, 202
Program search, 142
Program synthesis, 32
Proteomics, 92

R
Redundancy, 234
Relational model, 180
Replication, 169, 203, 206, 257
Representation, 5, 82, 142, 176, 210
RMSE, 33
Robustness, 145, 234

S
Scalability, 144
Self-reference, 183
Simulation, 2, 58, 84, 174
Statistical significance, 23, 203
Symbolic mediation, 178
Symbolic regression, 6, 31, 32, 97

T
TensorFlow, 122, 130
TensorGP, 122–124
TGPGAN, 119, 128, 130, 133
Thermodynamics, 3, 170
TPU, 118
Trustworthiness, 82, 147

U
Unity3D, 58

V
Visualization, 61, 85, 97, 212
Visual Turing test, 142

Z
Zoetic systems, 183

Printed in the United States
by Baker & Taylor Publisher Services